Mysteries of Life

Great Mysteries

Aldus Books London

Mysteries of Life

by Stuart Holroyd

Series Coordinator: John Mason
Art Director: Grahame Dudley
Designer: Julia Jones
Editor: Nina Shandloff
Research: Sarah Waters
Series Consultant: Beppie Harrison

SBN 490 004334
© 1979 Aldus Books Limited London
First published in the United Kingdom
in 1979 by Aldus Books Limited
17 Conway Street, London W1P 6BS

Printed and bound in Hong Kong
by Leefung-Asco Printers Ltd.

Introduction

What exactly is the nature of life? This age-old question has recently been revived by scientists in the light of new discoveries, and it sets the author off on a fascinating journey through the mysteries of life. All living organisms are based upon that tiny sac of life, the cell. How, he asks, are cells constructed? Are they simply made of complex chemicals or do they contain some special ingredient, some mysterious force that makes life possible? Other chapters explore equally mysterious aspects of life, such as sensory perception, human behaviour patterns, body rhythms, and the oddities of inheritance. Just how does the human brain work? Is there a limit to its potential power? And now that men have taken the first steps into space, what are the chances of other life forms sharing our galaxy? Finally, the author looks at the apparently inevitable process of ageing before facing up to the greatest of life's mysteries – death itself. Can even doctors be sure when a human life is at an end? The answers to these questions and many more provide the author with material for a wide-ranging exploration of the many unsolved mysteries of human existence.

Contents

Chapter 1
What is Life?

How do we determine what is alive and what is not? What characteristics does man share with bacteria or viruses that could be used to define life? The discovery of the microscope revealed for the first time the existence of otherwise invisible creatures who share our world. Now scientists are delving deep into the cell itself, discovering how information is passed along through the generations. L-fields in humans have been discovered which may link our body chemistry with the movements of the stars and moon. Can these be part of the mysterious "vital principle" of life, sought by man through the ages?

Big fleas, observed the early 18th-century British poet Jonathan Swift, have little fleas upon their backs to bite 'em, and little fleas have littler ones, and so on *ad infinitum*. Although today we may find Swift's lines in collections of comic verse, in his day they expressed a revolutionary scientific observation. It was a Dutch lens-maker named Antony van Leeuwenhoek who constructed a simple microscope and, about 1675, first revealed to the world the wonders of microbiology. Leeuwenhoek wrote a long essay on the flea, tracing the history of its changing forms from its first emergence from the egg, and asserting that "this minute and despised creature" was "endowed with as great perfection in its kind as any large animal." He observed that in its early stages the flea is sometimes attacked and fed upon by other, even smaller parasites, and perhaps he thought that this information would inspire sympathy for the poor despised creature. All we know that it inspired was Swift's verse, but Leeuwenhoek with his microscope made as great and revolutionary an impact on scientific thought as Galileo had made a generation or two before with his telescope.

Life, the philosophers of the ancient world had taught, is generated spontaneously out of *inanimate*—that is, nonliving—matter. A belief common to the ancient Egyptian, Mesopotamian, and Hebrew religions was that man was made of clay, and had had the breath of life breathed into him by a god. Until Leeuwenhoek produced his microscope, nobody had suspected

Opposite: medieval painting illustrating the Old Testament view of the origin of life. God, a patriarchal figure, is surrounded by the multiplicity of animals which are the fruits of his creation as two angels look on.

Right: contemporary engraving of Antony
van Leeuwenhoek (1632–1723), the Dutch
lens-maker who first revealed the existence
of microorganisms too small to be seen by
the naked eye. He spent years studying the
structure of minute parts of living things
such as red blood corpuscles, capillaries,
sperm cells, teeth, and muscle. Many of his
discoveries and observations were published
in the *Philosophical Transactions* of the
Royal Society in London, of which he was
elected a member in 1680.

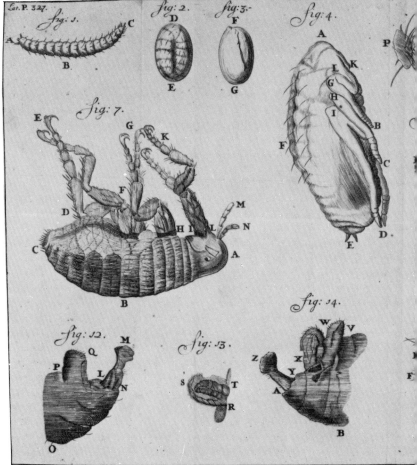

the existence of the swarming world of microorganisms, and the belief that the lower forms of life were "bred from corruption," meaning they arose from rotting substances, was universal. At first, Leeuwenhoek's discoveries seemed to confirm the belief in spontaneous generation, for anyone could now observe how microscopic life forms would soon appear in a food source like milk or broth if it were first sterilized and then kept warm for a while. But Leeuwenhoek argued that the seeds of these life forms came out of the air. There was a lively controversy over the question for nearly two centuries, until the great French scientist Louis Pasteur conclusively proved the pioneer Dutch microscopist right. The very air we breathe, Pasteur demonstrated, is full of organic life, and the ancient belief in spontaneous generation was an unfounded guess made in a time that lacked the instruments for precise observation.

But Pasteur's observations did not answer the fundamental questions: what is life? how did it originate? where precisely do we draw the line that separates the animate from the inanimate? Indeed even now, a century later, these questions are still debatable, although scientific technology has provided incredibly powerful instruments for exploring the nature of living things. In the 20th century what are known as the life sciences have produced a knowledge explosion as great as that in the physical sciences such as chemistry, physics, or astronomy, but still the

Spontaneous Generation?

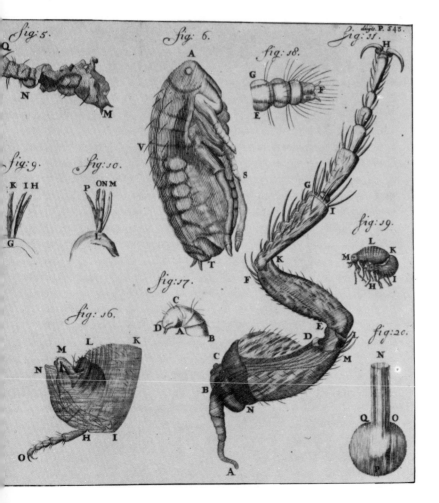

Left: the development and anatomy of the flea as drawn by Leeuwenhoek for his *Opera Omnia*, published in 1722. He reared his fleas from eggs, feeding the larvae on dead flies "which they devoured eagerly." Figure 7 (middle left) shows the full-grown flea; figure 12 (bottom left) the genital organ of the male flea; figure 10 (center) the parts of the mouth; and figure 15 (bottom center) the female genital organ.

mystery of life remains essentially unsolved.

The odd thing about this situation is that the differences between living and nonliving things seem obvious, and in our everyday behavior we all continually distinguish between the two. But the grounds on which we distinguish do not stand up to close examination, particularly when we start investigating the border areas. For example, shall we say that living things are characterized by the fact that they grow? But so do clouds, and according to modern astronomers so do stars, out of the amalgamation of dust in space. In his science fiction novel *The Black Cloud*, published in 1960, the British astronomer Sir Fred Hoyle ponders the question of what life is and plausibly imagines a cloud that is an organic, intelligent creature. Very well, that is fiction. But to exclude nonfictional clouds from the category of living things shall we define growth as taking material from the environment and reorganizing or shaping it to a predetermined pattern? Yes, but crystals do that. Are they alive? What about another tack—the ability to reproduce oneself? Is that not a characteristic only of living things? Maybe, but what about a virus? When a virus invades a bacterial cell it can multiply 100 times in minutes, using material taken from the cell itself. So is it alive, and are the viruses it produces, though not from its own substance, living organisms? It is difficult to exclude them from the category, but on the other hand those same viruses in their inactive phase are indistinguishable from crystals, and like crystals can be ground to a powder. Anyone can distinguish an elephant from a rock, but when science probes the frontier areas between life and nonlife it finds itself entangled in uncertainties.

Perhaps light could be thrown on the problem of what life is by considering how it originated on earth. Pioneers such as Leeuwenhoek and Pasteur showed that spontaneous generation of life does not occur, but their work did not prove that it had not occurred at some time in the depths of the past. Religions taught that life had been created from inanimate matter by a supernatural event of creation. A widely held alternative theory, which was supported by several prominent 19th-century scientists, held that temporarily inactive seeds of life were carried to earth from elsewhere in the universe as "passengers" on meteorites. Conventional evolution theory, on the other hand, maintained that life originated by a slow process of ordinary chemical reactions, while another school of evolutionary thought held that it originated in one extraordinary event, which might be highly improbable but nevertheless was statistically bound to occur given enough time. Many of these ideas were impossible to test, but as the science of biochemistry developed it produced a body of facts about the chemical characteristics of living things. They are basically composed of chemical compounds of just 12 of the natural elements—hydrogen, carbon, nitrogen, oxygen, sodium, magnesium, phosphorus, sulfur, chlorine, potassium, calcium, and iron. Carbon is outstanding among the elements for its tendency to form complex bonds with others, and accordingly it is the main chemical component of living things.

The first organic chemical compound to be formed in a laboratory was *urea*, which occurs in urine and other body fluids. The synthesis was accomplished by the German chemist Friedrich

Below: Louis Pasteur (1822-95), who demonstrated that the air is full of organic matter and thus proved Leeuwenhoek correct. Pasteur's most important contributions to microbiology were his confirmation, through work with cattle anthrax, chicken cholera, and rabies, that germs cause disease and his vaccination methods for preventive treatment.

Wöhler in 1828. Before that time, chemists had considered the difference between organic and inorganic compounds to be due to the action of an unspecified "vital principle" in living matter, and laboratory recreation of substances found in and produced by living organisms was believed impossible. But Wöhler's breakthrough led others into this area of experimentation, and by the end of the 19th century a wide range of organic chemicals had been produced in the laboratory.

To explain how the 12 natural elements gradually combined to form the complex molecules essential to life, the scientists J. B. S. Haldane in England and A. I. Oparin in Russia independently produced similar theories in the 1920s. The Haldane-Oparin theory suggested the existence of a "hot, dilute soup" of chemicals, sloshing about on the earth's surface at the dawn of time and going through a series of chemical reactions. Under normal circumstances such a mixture would eventually balance itself out or settle in a state of rest. The theory held, however, that circumstances in the environment of the primitive earth were not as we know them today. Instead of oxygen, the earth's atmosphere contained primarily carbon dioxide, according to Haldane, or perhaps methane, according to Oparin. Due to the low concentration of oxygen, there would have been no ozone layer in primitive times. In our atmosphere, this layer protects us from

What Exactly Do We Mean by Life?

Left: the German chemist Friedrich Wöhler (1800–82), the first to synthesize an organic chemical compound. His creation of urea in the laboratory helped to discredit the doctrine of vitalism, which claimed that the existence of a "vital principle" distinguished organic from inorganic matter. This, it was believed, would prevent scientists from ever creating life artificially.

the burning effects of most of the ultraviolet radiation given off by the sun. At the same time, the primordial soup would have been kept in a state of imbalance by a combination of volcanic action, ocean currents, rivers pouring fresh minerals into the mix, and deluges of carbon and nitrogen compounds from the atmosphere. Eventually, through the action of ultraviolet radiation on the chemical soup, the first organic molecules would have been formed.

The whole process had in fact been simulated in the laboratory, where ultraviolet radiation was found to act upon a mixture of water, carbon dioxide, and ammonia to form organic compounds, including the compounds basic to all life. Experiments done in the early 1940s, after the Haldane-Oparin theory was published, established that such reactions might also have come about through electrical charging of the primordial soup, for instance by lightning.

There is a big gap between simple organic compounds and the giant protein molecules of living organisms. Protein molecules are composed of chains of basic "building blocks," the *amino acids*, in various combinations. Amino acids make up a special class of organic compounds which are fundamental to life. For example, the insulin molecule consists of two chains of amino acids, one composed of 21 different amino acids and the other of 30. Protein molecules have different properties, or are of different kinds, depending on the way in which their amino acids are

Below: a painting showing the agitation of the primordial "soup" which is believed to have produced the first organic molecules leading to the emergence of life. The illustration is of a Pre-Scourian volcano erupting in the sea—named after the Scoury Dikes in Scotland.

Basic "Building Blocks" of Life

Left: a model of the lysozyme molecule, showing the complex structure of protein molecules. The yellow balls scattered throughout the model represent disulfide bridges; the terminal amino end is marked by the two blue balls; and the six red balls indicate active acid groups.

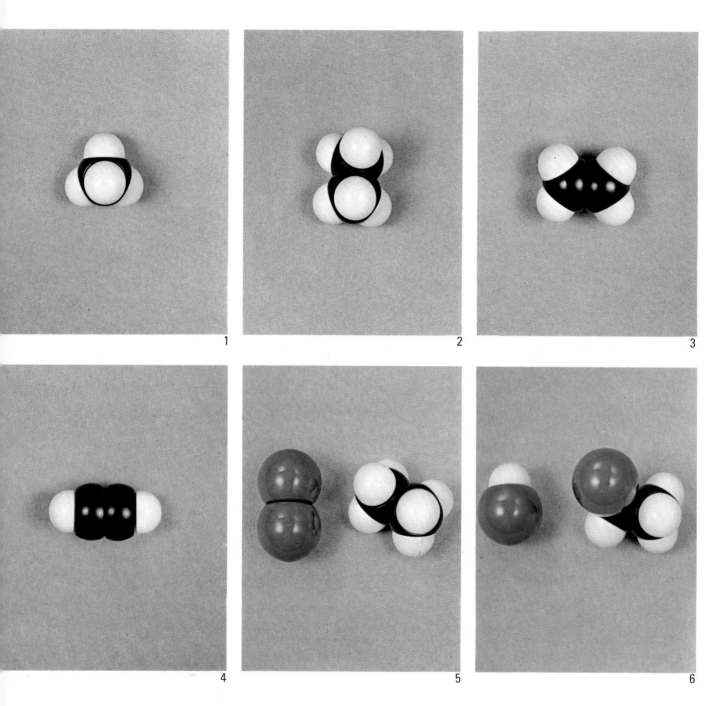

The carbon atom is the basic building block from which organic chemicals are formed. Above: a carbon atom (black) united with four hydrogen atoms (white) to form methane (1). Two carbon atoms and six of hydrogen make ethane (2); ethylene is composed of two atoms of carbon and four of hydrogen (3); two carbon and two hydrogen atoms form acetylene (4). Atoms of other elements can replace hydrogen in some compounds, such as the reaction (5) of ethane with chlorine (green balls) to form ethyl chloride and hydrogen chloride (6).

located in relation to each other. In chain A of the insulin molecule there are 280 million million possible permutations of the amino acids, and in chain B there are 510 million million million million. It took the British biochemist Frederick Sanger and his co-workers eight years to work out the precise permutations that determined the structure of the insulin molecule, an achievement for which Sanger was awarded a Nobel Prize in 1958. Taking this into account, as well as the fact that the simplest of the indisputably living organisms—the bacterial cell—consists of some 5000 different kinds of protein, the staggering complexity of the chemistry of life becomes apparent.

The next question is how the giant molecules could possibly have formed in the primordial soup. There are a number of theories to account for their formation, including the two extremes of random chance and divine intervention, but they are all unproven. All we know for certain is that life, in the form of the protein molecule, must eventually have gelled in the "soup." Life then developed into two distinct branches: plants and animals. Plants, which obtain their energy from light and from inorganic substances, must have been the first form of life to appear. Then probably followed an intermediate form, algae, and then the most primitive animals, floating in their source of nourishment, would have developed ways to absorb food and to reproduce themselves. Eventually, however, the easily-incorporated nutrients would have been exhausted. Life would have become subject to the obligation that in future would always apply to it: adapt or perish. Thus would have begun the evolutionary process, which the British naturalist Charles Darwin defined in 1859 as "the survival of the fittest." The famous phrase is unfortunately misleading, because "the fittest" is generally taken to mean the strongest and healthiest, and hence the conquerors, whereas in fact it meant that the organisms best

A Staggering Complexity

Below: ethyl alcohol (7) is created when ethyl chloride reacts with sodium hydroxide (NaOH), and the addition of oxygen produces water (H_2O) and acetaldehyde (8). Complex molecules such as aspirin (9) are constructed by a series of chemical reactions like this.

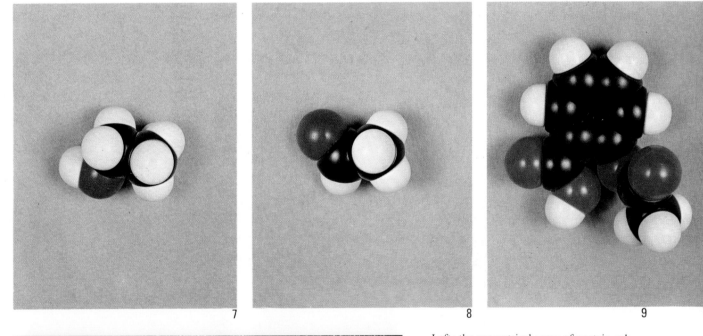

7   8   9

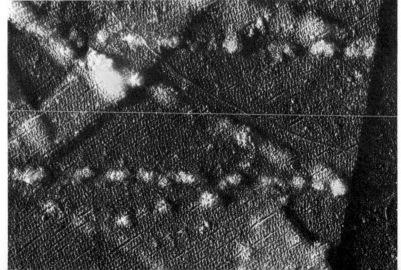

Left: the geometrical array of protein subunits in the outer layer of a *Distigma* cell as photographed through an electron microscope.

Simple forms of plant life are likely to have
evolved early in the chain of development
which began with the primordial "soup."
Right: *Spirogyra* with a water fern (Azolla)
and Lemna.
Below: fertile spikes of the great horsetail,
Equisetum telmateia.
Below right: liverwort and reproductive
archegonia, *Marchanta polymorpha*.

adapted and most adaptable to their environment survived. The survivors of the crisis caused by the initial dwindling of the food supply would have been those organisms which developed digestive mechanisms and new ways of obtaining their own nourishment—the foragers and predators.

Consideration of origins throws some light on the question of the nature of life, at least its nature on this planet. Life as we know it on earth depends on complex protein molecules, which must be capable of forming themselves into biological structures that interact with their environment. They must also be capable of reproducing themselves.

And life also depends on death. This is a key point. The forager or predator has to obtain protein nourishment by absorbing other living organisms. A basic characteristic of all forms of life is the ability to continually renew their own biological structures by processing organic material. As the British biophysicist Joseph Hoffman has written, "If it were not for the fact that life preys on life, all the organic material of the world would probably end up in a static group of living things. There would

Life Depends Upon Death

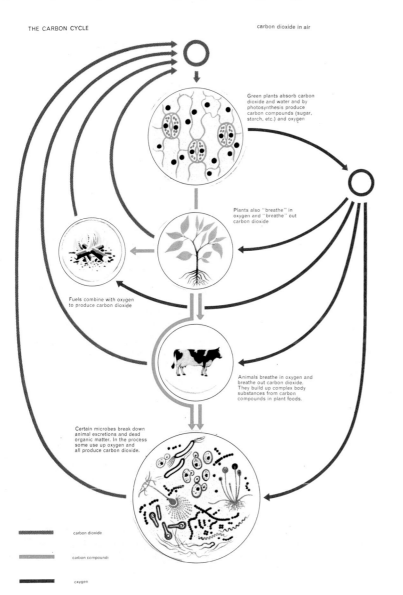

THE CARBON CYCLE

carbon dioxide in air

Green plants absorb carbon dioxide and water and by photosynthesis produce carbon compounds (sugar, starch, etc.) and oxygen

Plants also "breathe" in oxygen and "breathe" out carbon dioxide

Fuels combine with oxygen to produce carbon dioxide

Animals breathe in oxygen and breathe out carbon dioxide. They build up complex body substances from carbon compounds in plant foods.

Certain microbes break down animal excretions and dead organic matter. In the process some use up oxygen and all produce carbon dioxide.

carbon dioxide

carbon compounds

oxygen

Left: the carbon cycle. Photosynthesis by green plants uses the carbon dioxide in the air. Plant and animal respiration, fuel-burning, and the breakdown of animal wastes replace it, so that the cycle of life and death reprocesses the essential stuff of life.

The Infinite Chain of Life

be no food available. This static condition of things would be a dead end for all living things. They would have to go back to living on soil, water, and the sun's energy, like algae. This stasis [stagnation] is side-stepped: the living process has somehow arranged that as soon as a living animal looks good it will be devoured by another one which can in turn become better looking yet, and so on *ad infinitum*. The net result is that things keep changing and moving incessantly as long as life exists. The standstill is averted by death."

In the Hindu scripture the *Bhagavad-Gita* the god Vishnu reveals himself to Prince Arjuna in both his aspects—as the Creator in all his splendor and as the Destroyer in all his terror. Hindu thought has long recognized the dependence of life upon death. This may be because in the climate of India the processes of luxuriant growth and rapidly rotting death are plain to see. In many other parts of the world from time to time a sudden proliferation of one form of life makes people aware how dependent life itself is on a system of checks and balances between living things. Death plays a major part in this system. Plagues of locusts that can devour vast crops are still a hazard to agriculture in some

Right: Kalighat painting, dating from around 1860, of Hari-Hara, combined form of the Hindu gods Vishnu and Siva. Vishnu is the protective aspect while Siva fills the role of destroyer. Hindu thought considers the seemingly conflicting tendencies to be inseparable in the framework of our universe, because neither can exist without the other. Germination creates the tree, destroys the seed, and preserves the species; the carpenter creates the table, destroys the tree, and preserves the wood. That which has a beginning must, according to Hindu logic, have an end.

Left: living things die, are broken down, and then reappear in the life cycle in another form. Here snails are feeding on dead beech leaves—snails which themselves may be devoured later by birds or other predators.

parts of the world. When the Mormons first settled in Salt Lake City in 1847, swarms of crickets attacked their crops, and but for the intervention of a huge flock of seagulls that flew in all the way from the Pacific and devoured the crickets, the settlement would have been destroyed. More recently, at Okonto, Wisconsin in 1952, freak climatic conditions produced a plague of millions of large frogs. When the particular conditions favorable to one form of life prevail, and that form breeds beyond the capacity of its natural predators to control it, theoretically it could eventually take over the planet. Such an upset of the interdependent balance of life forms, or ecology, of our planet has been imagined by more than one science fiction writer.

Every individual life has to end, and the cells and molecules of every individual body have to be broken down and returned to the reservoir of organic material from which new life continually arises. This is the inescapable law of life. The molecules of life are passed from host to host, and once matter gets into the chain of life and has been part of a living thing it acquires a power to enhance the growth of other living things, a power which is not even destroyed by burning. Wood ash, for example, makes an excellent fertilizer.

This chain of organic activity implies that the basic stuff of life and the processes that sustain it are the same in all living things. But cells perform numerous different functions in living bodies. For instance, a brain cell works quite differently from a liver cell, although these functions are not chemically specified. Malfunctions of brain or liver cells cannot be cured by eating someone's or some animal's brain or liver. So the question arises— how do cells, which are composed of collections of giant protein molecules, and which in their turn are made up of chains of amino acids, become adapted to their specific biological functions? How, in other words, do they know what to do?

Just putting the question in this way points to the very basic

Above: the familiar sequence of animals feeding on plants is here reversed: an insect is trapped on the sticky projections of a sundew leaf, *Drosera intermedia*.

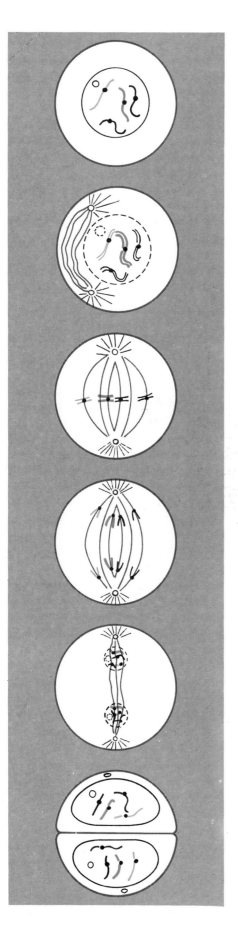

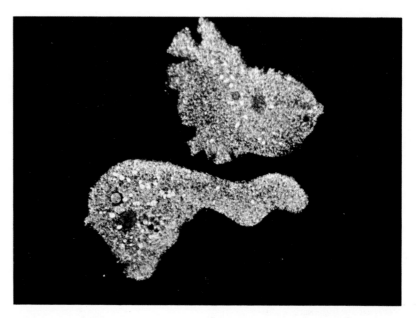

difference between animate and inanimate matter. Living matter depends for its existence upon *information*. The Second Law of Thermodynamics, one of the four fundamental laws of physics, states that physical systems inevitably tend towards a state of disorganization. Their parts are subject to *entropy*—that is, the process of becoming more similar. Biological (animate) systems are distinguished from physical (inanimate) ones by the fact that they reverse this tendency. They tend towards increased differentiation and therefore increased order. Systems that depend only upon a flow of energy eventually run down, but systems in which a flow of energy is combined with a flow of information are self-sustaining and capable of complex adaptations. The giant protein molecules have an information component. It is this that distinguishes them from inorganic matter.

An amoeba, a single-celled organism, normally divides every 48 hours, but if subjected to a burst of X-rays it starts dividing erratically, in a time period which may be as short as 12 hours or as long as 90 hours. What has happened is that the radiation has jammed or disrupted the cells' information-processing mechanism. Something similar seems to happen with developing human cells over a period of time. Eventually they reach what is known as the "Hayflick limit." L. Hayflick of the Wistar Institute in Philadelphia discovered in the 1960s that a culture of such cells under test tube conditions will continue to multiply only for about 50 generations. Cells which originated as clearly recognizable human body cells eventually produced offspring that seemed to "forget" what they were supposed to be. They lost their information component, became shapeless and anonymous, and soon after crossing the Hayflick limit they died.

The maintenance of internal order in the cell is the work of a class of protein molecules known as *enzymes*. Enzyme activity organizes the processes of chemical breakdown and synthesis that go on all the time in the living cell. In the words of the British biochemist Stephen Black, enzymes are the "marriage brokers of the living chemistry." They bond similar molecules together and assemble them into the substance of the body. They are thus the information-carriers of the cell, the components whose specific actions contribute to the fulfillment of the cell's overall purpose. It is their ceaseless activity that combats the trend towards increasingly random molecular movements and eventual breakdown of the cell.

But can information-processing in the cell, and in the bonding of cells into body substance, be explained entirely through chemical activity? This is a question upon which biochemists and biophysicists cannot agree. Black appears to support the view that it can, but Hoffman expresses an alternative view. He puts forward what he calls the "template hypothesis." A template is a blueprint. "In each of the billions of trillions of living cells in the world today," Hoffman writes, "there is a pattern, or guide, or mold, or working plan. This is an immortal molecular template because it has been reproducing itself incessantly since the beginning of life over two billion years ago." And he sees the changes of form which affect organic matter as a contest between templates: "The cycle of transformation from one kind of living unit to another is repeated billions of trillions of times each day. Matter in the form of atoms in molecules moves from one pattern to another, or, we should say, goes from the influence of one template into that of another. Literally all of the organic material spread so tenuously over the earth's surface is being sought for by the templates in living things. And while matter is moving from one to another the templates themselves undergo changes."

Another of Hayflick's observations would appear to support the template hypothesis. When a cell is approaching the Hayflick limit and begins to lose "memory" of its role and identity, it can be brought back to health and normality if restored to its original body. Being brought back into proximity to its own kind enables it to recover the information it had lost and thereby revitalize itself.

The reason many biophysicists do not like the template hypothesis is that it seems to bring back the indeterminate something,

The "Hayflick Limit" of Cells

Opposite (left): stages of cell division, or mitosis. First the membrane around the nucleus disappears while each chromosome (only two pairs are shown) splits in half. Spindle fibers develop in the surrounding cytoplasm, and these help to separate the duplicate chromosomes, which then move to opposite sides of the cell. A new nuclear membrane forms around each group of chromosomes, and finally the contracting cell membrane pinches through the middle of the cell, resulting in two cells which are exactly alike.

Opposite (right): two amoebas, single-celled organisms. The dark spots are contractile vacuoles, cavities which regularly contract to discharge fluid from the cell body.

Below: the processes of chemical breakdown (left) and synthesis (right). Digestive enzymes consist solely of protein. They split a substance (*substrate*) into smaller parts through the intense chemical activity of the enzyme's surface. Other enzymes often require a mineral or vitamin coenzyme in order to assemble small molecules of substrate into larger ones.

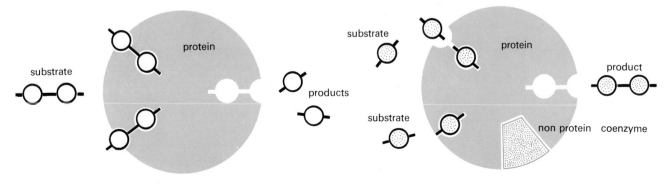

digestive enzyme   non digestive enzyme

The L-Field

the "vital principle" of the chemists of old. The difficulty may, however, have been solved by Professor Harold Saxton Burr of Yale University, who believes he has discovered the nature of the organizing principle in living systems. In 1971 he presented his findings and his arguments in the book *Blueprint for Immortality*, which he subtitled *The Electric Patterns of Life*.

The fact that all living systems have electrical characteristics has long been known. Electrical measurements of brain, heart, and glandular activity have been increasingly used in recent years, both in the diagnosis and treatment of disease. But, Burr argues, no general theory covering these electrical properties has been developed, because modern biology has been biased towards the purely chemical interpretation of life processes. "Chemistry rests upon a discontinuous, atomic conception of nature" and focuses its attention on separate "entities rather than upon structure, and on constituent elements rather than on the whole." He puts forward the general theory of the electrodynamics of life that he maintains has hitherto been lacking.

"It has usually been assumed," Burr writes, "that all the changes in the electrical properties of a living system are the consequence of biological activity. But it is our hunch that a primary electrical field in the living system is responsible." He calls this the "field of life," or L-field for short, and to explain it he uses a simple comparison. When iron filings are scattered on a card held over a magnet, they arrange themselves in a pattern corresponding to the "lines of force" of the magnetic field. You can keep changing the iron filings on the card, throwing them away and scattering a new lot, but the pattern they fall into never changes. "Something like this—though infinitely more complicated—happens in the human body," Burr continues. "Its molecules and cells are constantly being torn apart and rebuilt with fresh material from the food we eat. But, thanks to the controlling L-field, the new molecules and cells are rebuilt as before and arrange themselves in the same pattern as the old." When we reflect that to accomplish this feat every amino acid in every molecule must be in its precise place, the L-field might well be the most marvelous of life's mechanisms.

Burr's claim is not a hypothesis, but is based upon 30 years of research. Over that period, he wrote, "almost every form of living organism has been studied, some of them quite cursorily and others in more detail, from bacteria up to and including man. And so far as our present information goes, there is unequivocal evidence that wherever there is life there are electrical properties." One of his first discoveries was that the L-fields of human beings show voltage level changes which recur at regular intervals. He found that it was possible to pinpoint the monthly occurrence of ovulation in females by monitoring their L-fields and looking for a sharp rise in the voltage. L-field measurements were also found to be helpful in locating malignancies in the body, and in measuring how quickly internal wounds were healing after operations. When measurements were taken of seeds it could be predicted how strong and healthy the future plants would be. These are all practical applications of the technology Burr had developed to measure L-fields, and they proved the reality of the discovery. But Burr saw beyond these practical applications to more signi-

Left: portrait photograph of Harold Saxton Burr of Yale University. In his book *Blueprint for Immortality* he puts forward a general theory of the electrodynamics of life, which involves, he claims, an electric "field of life" or L-field.

ficant, indeed revolutionary, theoretical implications. He saw that L-fields are at the same time the organizing principle in both the protein molecule and the individual body cell as well as our "antennae to the universe."

This last phrase requires some clarification. Burr discovered that living organisms respond to many environmental influences through changes in their L-fields. Burr recorded L-field measurements for a large maple tree in his garden, which he kept wired up over a period of years, and he found voltage changes which corresponded with lunar cycles and sunspots. Human beings, he suggested, must also be subject to such influences. The age-old belief of poets and mystics—that all of life is one, continuously involved in complex interactions—is borne out by Burr's discovery.

From Leeuwenhoek, who discovered the microcosmic abundance of life, to Burr, who discovered its uniting force and universal interrelations, the life sciences have taken immense strides in 300 years. But for all this progress, there are still mysteries that remain to be solved.

Below: in 1966–7 a follower of Professor Burr, Ralph Markson, undertook a study to determine whether the geophysical environment was related to the electric potential in one form of living system—a tree. A typical day's record compares the hour-by-hour electrical potential of a dying maple tree, a healthy elm, and the surrounding air and earth. The different potentials appear to be largely "in phase," lending weight to Burr's theory that living things respond electrically to their environment.

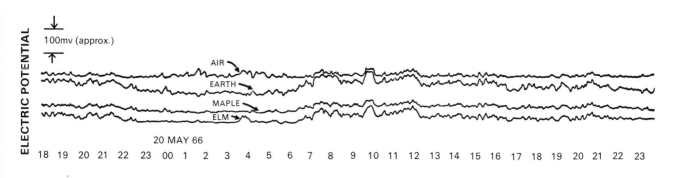

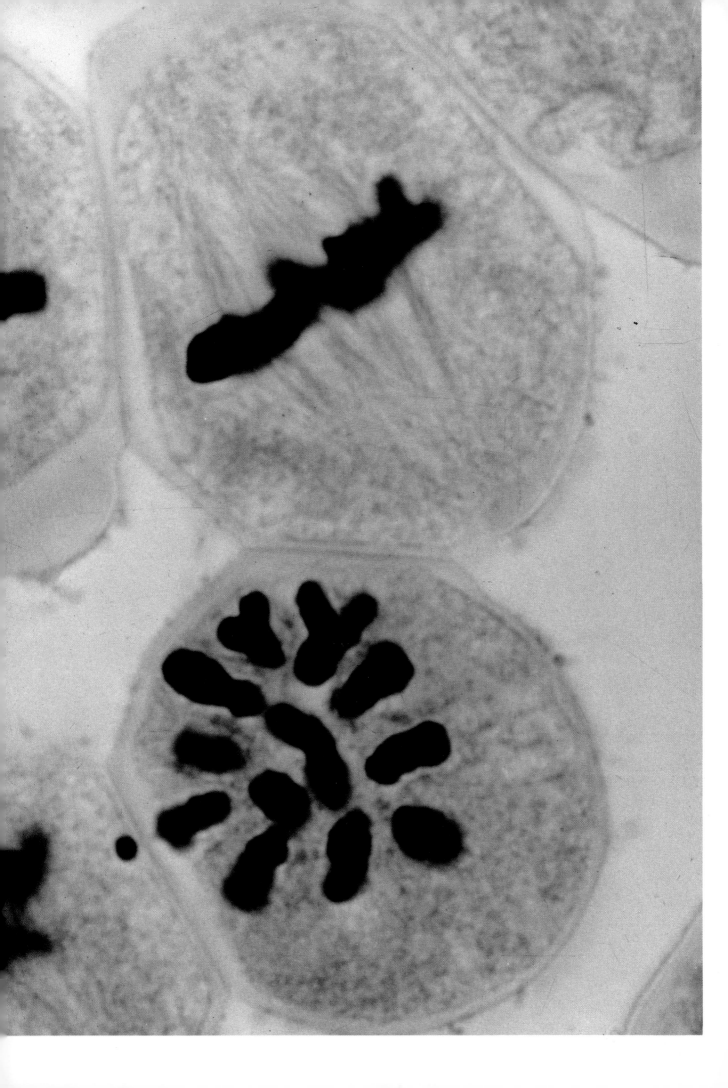

Chapter 2
Secrets of the Cell

Why do children look like their parents—and why do they sometimes not? How do the studies of common pea plants made by an obscure Moravian monk over a century ago relate to human genetics? Since then scientists have made great strides in determining the make-up of the genes that carry our heredity—including an exciting race to decipher the complex DNA molecule. And what of the mystery of how each single cell in the human body "knows" what it is meant to do—can a cell forget? How do brain cells work? Although some of these secrets have been discovered, many remain mysteries which science cannot yet explain.

One evening in February, 1865 the 40 members of the Society for the Study of Natural Science met in a room at the schoolhouse in the little town of Brunn in Moravia (now Czechoslovakia) to hear a lecture by a priest from the nearby monastery. His subject was "Experiments in Plant Hybridization." The circumstances could not have seemed likely for the first announcement of an epoch-making scientific breakthrough. Nor did the audience in fact appreciate what happened that evening, for the Society's minutes recorded that there were no questions. Indeed, it was not until the 20th century that the significance of the work done by that obscure priest, Gregor Mendel, was understood. He was then acclaimed as the man who had, without a microscope, probed the mysteries of the cell.

The modern science of genetics is based on Gregor Mendel's conclusions. The paper he had read to the scientific amateurs of Brunn that winter evening in 1865 was simultaneously redis-covered quite independently by three distinguished botanists—De Vries of Holland, Correns of Germany, and Tschermak of Austria—in the year 1900. Gradually over the following decades the scientific community realized that the rules of heredity which the Moravian priest had formulated for plants applied to all living things, including man.

Progress in science depends on someone putting the right questions to nature. Scientists of the 19th century were mainly interested in the processes of change and evolution, and Mendel

Opposite: a photomicrograph taken during the process of mitosis, or cell division. The characteristics of the two daughter cells will be determined by instructions carried on the chromosomes, the small pieces of living material which have here been stained darker for easier identification.

Gregor Mendel, Father of Genetics

was alone among them in that he was interested in why things didn't change very much. He wanted to know what the laws of organic stability were. Experimenting in his monastery garden with different varieties of the common pea, Mendel systematically crossbred them. He separated out specific characteristics in the strains he developed, combining these characteristics in different ways and observing what happened to them as they were passed on through several generations. For instance, when he crossed a tall and a dwarf variety of pea he found that their offspring were all tall. These in their turn produced a generation of hybrids in which the characteristics were distributed in the ratio of three tall to one dwarf. Mendel's observation that characteristics were inherited according to very specific laws led him to the conclusion that in the reproductive cell there must exist distinct units—*genes*—which carry these characteristics. In Mendel's day it was universally believed that characteristics from both parents blended in the process of reproduction. His discovery that they do not blend implied that when two separate genes come together one characteristic emerges as *dominant*, possibly over several generations, and the other becomes *recessive*. The recessive one

Right: a plaque of Gregor Mendel (1822–84), based on a photograph taken during his lifetime. Considered the founder of the science of genetics, Mendel reasoned that certain basic laws governed the heredity of characteristics in living things. Luckily he chose to study traits determined by one pair of genes, since single characteristics can frequently result from a combination of several genes.

Opposite: results of an experiment in crossbreeding sweet peas, described in 1909 by William Bateson, an early and ardent follower of Mendel. The purpose was to determine whether a link exists between the genes which determine the color of the flower and those which control its shape. First a plant of the type called Emily Henderson (1), with a white, flat (or erect) flower, was crossed with a white, "hooded" type known as Blanche Burpee (2). The product—or first generation offspring, F_1— was an erect, two-toned purple flower called the Purple Invincible (3). When F_1 was self-fertilized, the various offspring obtained in the F_2 generation (4–11) were either purple (two-toned erect or single-color hooded); red (two-toned, erect only); or white (single-color, erect or hooded). Notice that while purple and white flowers can be either erect or hooded, the reds are always erect.

THEODOR CHARLEMONT

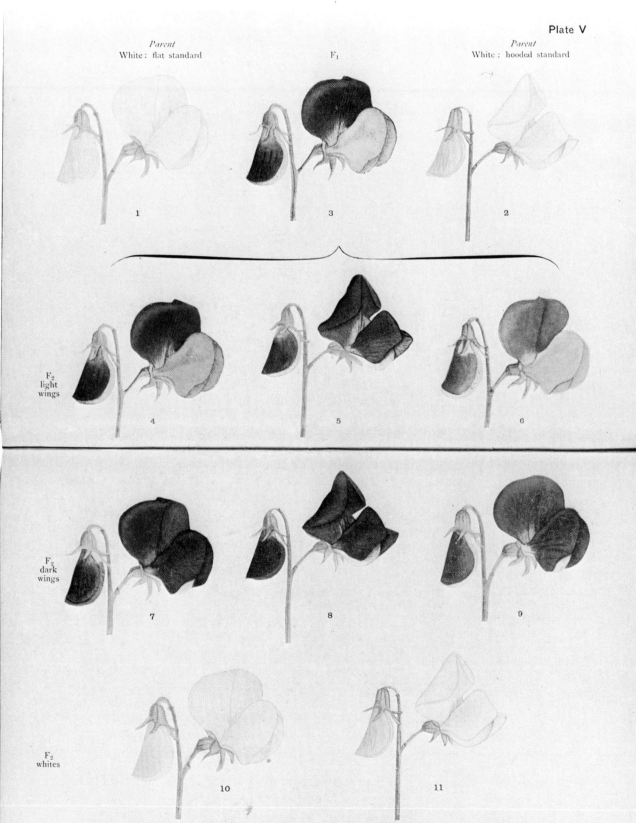

Plate V

1. Emily Henderson. 2. Blanche Burpee. 3. Purple Invincible, F₁. 4—11. The various F₂ types obtained by self-fertilising F₁. 4. Purple Invincible. 5. Duke of Westminster. 6. Painted Lady. 7—9. Corresponding dark winged types. 7. Purple, with purple wings. 8. Duke of Sutherland. 9. Miss Hunt. 10 and 11. F₂ whites. Notice that there is no *hooded* red.

Right: beekeeping was another of Mendel's interests. His beehives occupied a large but separate part of the monastery garden, away from the site of his experimental pea plants.

lies dormant but still retains its identity, and it may emerge with no loss of strength in a future generation. This information was tremendously important for two reasons. First, because it explained the continuing variety within a species (for if blending were the rule the trend would be away from extremes and towards undifferentiated sameness). Second, because it opened the way for scientific analysis of genes as the basic control units of life, if only by establishing that they must exist.

Throughout the last quarter of the 19th century many biologists were studying *chromosomes*, which are rod- or sometimes ribbon-shaped parts of the nucleus of a cell. They discovered that each organic species has a fixed number of them in its cells. In man the number is 46 (23 pairs) and in the mouse it is 40 (20 pairs), for example, but there is no orderly relationship between the complexity or degree of development of a species and its chromosome count. Every cell in the body has the same specified number of chromosomes with the exception of the reproductive cells—the egg and sperm in mammals—each of which has half the number so that when they join together the characteristic chromosome count is restored.

In 1908 a professor of zoology at Columbia University in New York, Thomas Hunt Morgan, developed an interest in Mendelian heredity theory and began experimenting with the common fruit fly. He found that in its immature larval stage the fly has giant chromosomes in the cells of its salivary glands, which are identical to those in its reproductive cells but hundreds of times larger. They were large enough, in fact, to be studied under a microscope, and when studied it was observed that they had a series of some 600 thin, flat strips along their lengths. These strips were found to contain factors which determined such hereditary qualities as hairiness, wing shape, or eye color. Mendel had suggested the existence of the gene as the carrier of a "unit-character," and Morgan's contribution was to identify and locate it as a part of the chromosome.

With the merging, or fusion, of reproductive cells, chromo-

The Chromosome

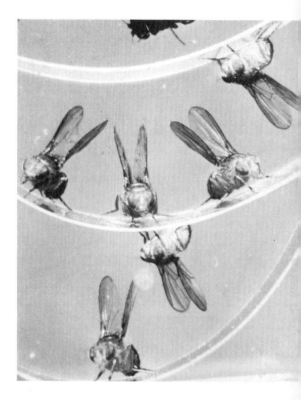

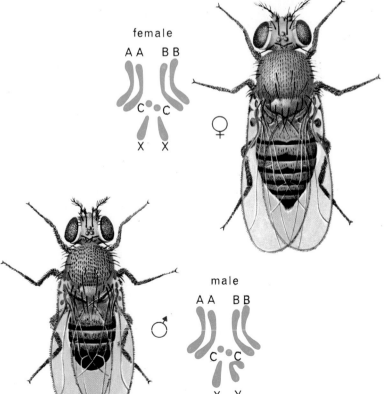

Above left: American biologist Thomas Hunt Morgan (1866–1945). Co-author of *Mechanism of Mendelian Heredity* (1915) and author of *The Theory of the Gene* (1926), Morgan's contributions to the "chromosome theory of heredity" gained him the Nobel Prize for Medicine in 1933.

Above: upheld wings denote mutated specimens of *Drosophila* (fruit fly), subjects of Morgan's experiments on chromosomes and heredity.

Left: the shape of the chromosomes of *Drosophila*. Female sex cells contain one chromosome of each pair—ABCX. Male sex cells can be denoted either ABCX or ABCY. Their characteristic shapes helped Thomas Hunt Morgan and his team to isolate and study their role in heredity.

somes pair off at random, and each of the resulting 23 pairs exchanges genetic material with its partner. This haphazard process of chromosome-pairing and genetic exchange in reproductive cells between the male and female sets guarantees the continuance of an infinite variety of human beings. We have all wondered at the vast number of different human faces in the world, and at wide differences in features and other characteristics between children of the same parents. The answer to this mystery, as the British zoologist Sir Peter Medawar wrote in the late 1950s, is that "the combinational varieties of known human genes outnumber all the people who are alive today, have ever lived or are likely ever to live."

It was not, however, until the 1940s that biochemists discovered what chromosomes consist of and began to understand how the genetic messages are coded in them. Working on a suggestion made by the British bacteriologist Dr. Fred Griffith, scientists at the Rockefeller Institute for Medical Research in New York found that hereditary instructions are conveyed in the structure of the deoxyribonucleic acid molecule, or DNA. This discovery, made in 1944 by a team led by Canadian bacteriologist Oswald Avery, started one of the most exciting races in scientific history. As the American scientist James Watson, one of the winners of the race, has frankly written, in the 1940s and early 1950s "DNA was still a mystery, up for grabs, and no one was sure who would get it and whether he would deserve it if it proved as exciting as we semi-secretly believed."

Right: Maurice Wilkins of King's College, London in 1962. Using X-ray crystallography, he and his colleagues were able to confirm the essential assumptions and predictions embodied in the Watson-Crick model of the DNA molecule. As one of the first to investigate DNA, he shared the Nobel Prize for its discovery with Francis Crick and James Watson.

Watson's book *The Double Helix*, subtitled *A Personal Account of the Discovery of the Structure of DNA*, tells the fascinating story of frontier scientific work, including all the elements of ambition, rivalry, backbiting, blunder, luck, and exuberant excitement that are all part of it. The following paragraphs are largely based on Watson's account.

After Avery and his team had made their discovery in New York, research on DNA was taken up by British biophysicist Maurice Wilkins at King's College, London, who made it his special preserve. An unwritten gentleman's rule existed among British scientists that one could stake out a research area and explore it at one's leisure, and that others wouldn't trespass on the territory. But it seemed to some people, particularly the brilliant Francis Crick at the Cavendish Laboratory at Cambridge, that Wilkins' approach was too leisurely. Crick wanted to work on DNA himself, and found the scientific protocol frustrating. He also feared that someone who was not bound by the protocol would unravel the structure of DNA before Wilkins did and before he himself could have a crack at the problem. His fear was increased when the great American Nobel laureate chemist, Linus Pauling of the California Institute of Technology, wrote to Wilkins and asked to see copies of his crystalline DNA X-ray photographs. Wilkins replied that he wanted to look more closely at the data before releasing his photographs. But it was clear that the formidable Pauling was on the scent of the most momentous discovery in modern biology.

Another man who was on the scent was James Watson. Fascinated by genetics since his undergraduate days, Watson heard Wilkins talk about his X-ray work on DNA at a scientific meeting in Italy in 1951, and he became determined to crack the DNA code. He tried to get Wilkins to take him on as a co-worker at King's College, but without success. The rebuff did not affect his enthusiasm—Watson got himself a job at the Cavendish Lab in Cambridge with the like-minded Francis Crick. "Finding some-

The Start of the DNA Race

Left: James Watson and his colleague Francis Crick with their double-helix molecular model of deoxyribonucleic acid. They were the winners of a dramatic race within the scientific world to solve the puzzle of the structure of DNA. The photograph was taken at the time of the Nobel Prize award in 1962.

False Trails!

Right: reconstruction of Crick and Watson's 1953 model of DNA. It incorporates the metal plates they used to represent the base nucleotides. Composed of two chains of nucleotides, the backbones run in opposite directions and are coiled around each other to form a double helix. The base nucleotides project into the center of the helix. Later detailed calculations showed that, although a few minor features required modification, the model essentially represents the actual structure of DNA.

one in the lab who knew that DNA was more important than proteins was real luck," he later wrote.

Analysis of the structure of protein molecules was the main interest of biochemists at this time, and Pauling himself had just published a series of papers that went a long way towards solving the structure of proteins. A protein *macromolecule* is a giant molecule, built up of amino acids in a chain formation. Pauling's discovery was that the structural arrangement, fundamental to the chemistry of all proteins, is helical—that is, the molecular chain is twisted as if around a pole. Pauling had built a model of his "alpha helix" which he dramatically unveiled with the announcement of his discovery, and Watson and Crick launched their attack on the DNA problem using the same technique of model-building. They soon found, however, that DNA was a much more complex molecule than a protein. X-ray photographs of it in its pure solid crystalline form indicated that it was thicker and probably contained more than one chain of chemical units known as *nucleotides*. How many chains it contained, what precisely the nucleotides were, and how many there were of them were unanswered questions. Knowing that Pauling was working on the problem, Watson and Crick approached Wilkins and suggested cooperation.

Wilkins was agreeable, but he had a co-worker, Rosy (Rosalind) Franklin, who was not, and she had the X-ray photographs and the expert knowledge of crystallography that were essential for work on the DNA problem. Rosy scorned the model-building approach, and when Watson and Crick put together their first model, which consisted of three chains of nucleotides twisted around an axis, she quickly pointed out a basic fault in their theoretical design which they had both missed. This failure resulted in Watson and Crick being moved from DNA research by their professor, Sir William Lawrence Bragg. Watson, however, managed to pursue his obsession by working on the tobacco mosaic virus (TMV), which had a ribonucleic acid (RNA) component. RNA is the "messenger" substance that carries the instructions of the DNA for building protein molecules. He hoped that "if we solved RNA we might also provide the vital clue to DNA." He had notable success in his work, and was able to establish that the structure of TMV was helical.

Linus Pauling's son Peter was a research student at Cambridge at this time. Watson and Crick were almost disappointed when they learned from him that his father had written in a letter that he had worked out a structure for DNA and would be publishing his findings soon. Their feelings turned to delight, though, when Pauling published his model and they found that he had made errors even more obvious than the ones they had made in their first model months before. Gleeful at the thought "that a giant had forgotten elementary college chemistry," and at the same time worried that when he discovered his mistake Pauling would devote all his intellectual energy to DNA, Watson and Crick were anxious to make the most of the start that they had on him. As a result they managed to persuade Bragg to approve their resuming work on DNA.

The temporarily-imposed halt on experimentation had not applied to thinking about the problems of DNA. Crick had been considering certain regularities in DNA chemistry observed by the American biochemist Erwin Chargaff at Columbia University. Chargaff had established that there were four basic types of nucleotide in DNA: adenine, thymine, cytocine, and guanine (A, T, C, and G). He had also found that the numbers of A and T molecules tended to balance and so did the numbers of G and C molecules. Crick deduced that there must be forces of attraction at work between A and T and between G and C, and he had a hunch that this might be a key to the structure of DNA. When his hunch was combined with Watson's that the DNA structure would be a double helix—two nucleotide chains twisted around each other—they were back in the business of model-building. And this time, in 1953, they got it right, as even Rosy Franklin and Linus Pauling had to admit. The race was over. Nine years later, in 1962, Francis Crick, James Watson, and Maurice Wilkins shared a Nobel Prize for unraveling the structure of the DNA molecule.

To look at, the Watson-Crick model is a jumble of tightly-packed, different colored balls (representing the different nucleo-

Below: Linus Pauling with one of his atomic models. His research on the structure of DNA at the California Institute of Technology put pressure on Watson and Crick and helped turn the search into the dramatic race it became.

Below: this triple-stranded helix as
proposed as the structure of DNA by Linus
Pauling in California in 1952. Pauling's
main error, due to inadequate information,
was in suggesting that the three helical
chains were held together by hydrogen
bonds between the phosphate groups at the
center of the helix.
Right: the "ball-and-spoke" model,
accurately showing one complete turn of the
double helical structure of DNA. Ten
nucleotide pairs form one complete turn of
the helix.

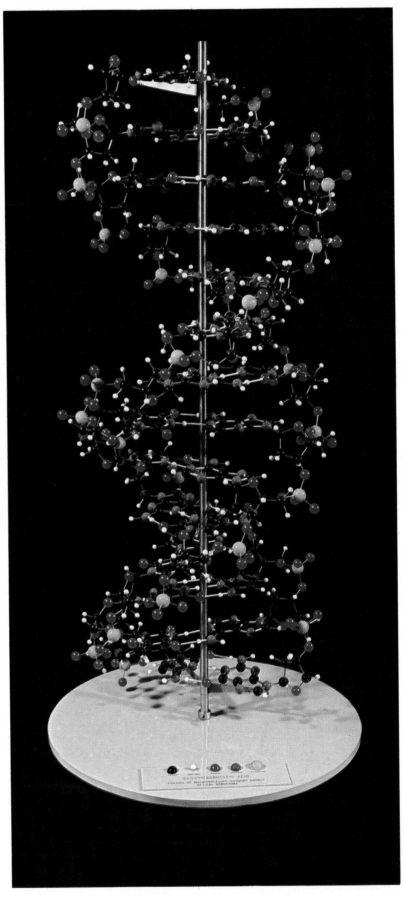

tides), composed of two strings twisted together. It is far from clear what it does and how it forms a code. We are told that this chain contains information and controls the functioning of the cell, but how it does these things is a mystery to the non-scientist, and indeed was a mystery to the experts at the time of the original breakthrough. The answers were only found through the work of many researchers over the next two decades.

A code is a system of correspondences between two different sequences of events. Say, for instance, I had four picture cards, of an apple, an orange, a pear, and a lemon. I could decide that a combination of any three in a particular order would stand for a letter, and could work out an alphabet on this basis. In the DNA code the letters are the four base nucleotides—A, T, G, and C. Each sequence of three of them on the chain spells out a "word," or a specification for a particular amino acid. As a DNA molecule can contain up to 10 million nucleotides in its two chains, a lot of "words" can be spelled out. The amino acid chains—or "sentences"—that are formed of these "three-letter words" specify particular proteins. The "sentences" are the genes. Each gene instructs the making of a specific enzyme, a special protein which in turn controls a particular chemical reaction within the cell. In order to produce a specific result—for instance, the manufacture of a hair-color pigment—the interactions of several genes and their associated enzymes may be required. And as different cells perform different functions, usually only a small section of the DNA chain will be used, just enough to specify and control a cell's particular function, and the rest of the chain will remain unused.

If the information-processing abilities of DNA are mind-boggling, even more so is the fact that under suitable conditions (and with the help of appropriate enzymes) it can reproduce

The Breakthrough

Left: the final accolade. At the 1962 Nobel presentation ceremony in Stockholm, Sweden are Professor Maurice Wilkins (far left), Dr. Francis Crick (third from left), and Professor James Watson (second from right).

Methods of Reproduction

Right: the nucleotide unit from which chromosomes are built, consisting of a phosphate, a sugar (deoxyribose), and a nitrogen compound, or base. The sugar-phosphate backbone never varies, but the base component can be any one of four possibilities: guanine (G), cytosine (C), adenine (A), and thymine (T). The nucleotide units are the individual "letters' of the DNA code.

Below right: many nucleotides bond together to form a polynucleotide thread, the "words" and "sentences" of the DNA code.

Far right: two long threads coiled around each other in a spiral form the double helix of the DNA molecule. The base components of the two threads are joined together across the spiral by hydrogen bonds, shown by dotted lines; guanine always pairs with cytosine and adenine with thymine.

Opposite (below): diagram of a one-celled plant (*Chlamydomonas*) showing the process of asexual reproduction, which involves the fission or splitting of the cell. The parent "disappears" in the sense that two "daughter cells" are the end result.

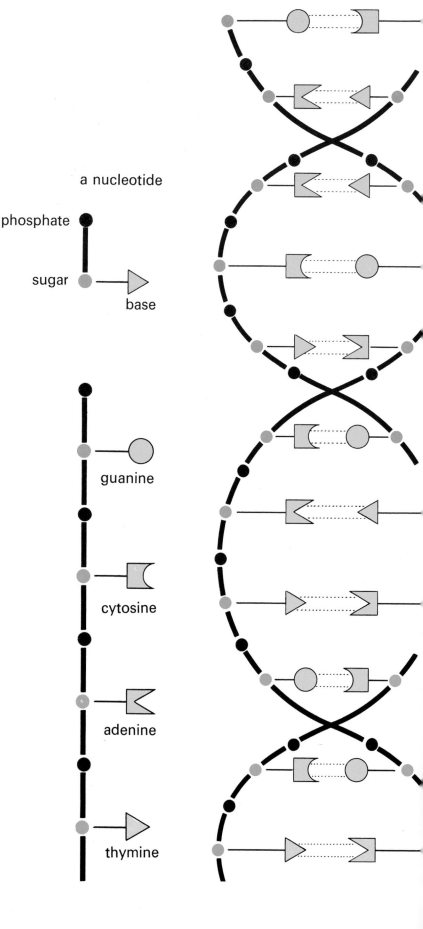

a nucleotide

phosphate

sugar — base

guanine

cytosine

adenine

thymine

itself. This process of self-reproduction occurs by cell division, called *fission* or splitting, which takes place at a rapid rate in the growth of any organism from its initial undeveloped stages. During cell division the chromosomes split in half, the rest of the nucleus divides and one full set of identical chromosome pieces reassembles in each one of the two newly-formed cells.

Some very simple organisms reproduce themselves entirely by division, but the method is not entirely satisfactory over the long term, because after several hundred reproductions the process tends to slow down and eventually stops altogether. Nature has devised sexual reproduction as a way of overcoming this difficulty. In human reproduction, as previously mentioned, two cells with only 23 chromosomes each merge, or fuse, to form one complete cell which has a full set of 46 chromosomes, half from each parent. One of the main advantages of sex, from an evolutionary point of view, is that it involves complex interactions between different groupings of genes. This insures strength and variety within the species.

Reproduction by fission—genetic duplication—or *fusion*—genetic exchange—are nature's ways. The possibility of a third way, "genetic engineering," has been seriously proposed by a number of scientists in recent years. In 1970 Professor F. C. Steward of Cornell University published a description of an experiment in which he had taken cells from a carrot root and produced from them a complete new plant. The process was called *cloning*, from a Greek word meaning "a throng." Cloning people—that is, producing replicas of any individual from a single cell—is conceivably possible. Since the DNA in any individual cell contains all the information necessary to specify a complete person—although most of that information is normally ignored to allow the cell to get on with its particular job—it is possible in theory to duplicate a person by a process of repeated

Above: asexual reproduction of a tulip bulb. The daughter bulbs, or *bulbils*, are developing around the side of the parent bulb.

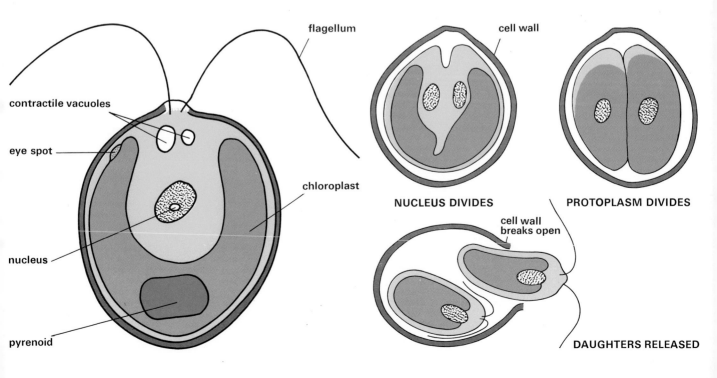

flagellum

contractile vacuoles

eye spot

chloroplast

nucleus

pyrenoid

cell wall

NUCLEUS DIVIDES

PROTOPLASM DIVIDES

cell wall breaks open

DAUGHTERS RELEASED

cell division beginning with one cell. Some sensational speculations have been made about the application to which the cloning
technique might be put, from the benefits of having 50 Einsteins
or Beethovens to the dangers of having 50 Hitlers or Stalins. But
the general opinion among most scientists today is that the technique will not be developed, although the ethical difficulties that
might prevent it are greater than the scientific ones.

To conclude our study of the cell, it is useful to take a close
look at one of the most interesting cells in the body. There are
two basic types of brain cell—*neurons* and *glial cells*. The brain
neuron is unique among human body cells because it is so highly
specialized that it will not reproduce itself. After the first months
of life we all have as many brain neurons as we will ever have;
indeed, when we have our full allowance they begin to die off at
a rate of tens of thousands daily. In fact, over the course of a lifetime we only lose on average about 3 percent of the total. When
neurons die off they are replaced by glial cells.

Right: the Italian physician Camillo Golgi
(1843–1926), who was awarded the Nobel
Prize for his work on the structure of the
nervous system. In 1883 he demonstrated
the structure of nerve cells and their
branches using his own silver nitrate method
of staining, a method which made possible
the development of modern neurology. He
was also known for his valuable
observations on malaria, pellagra, and
mental disease.

Below: a Golgi-stained section of human
brain tissue from his book *Opera Omnia*,
published in 1903.

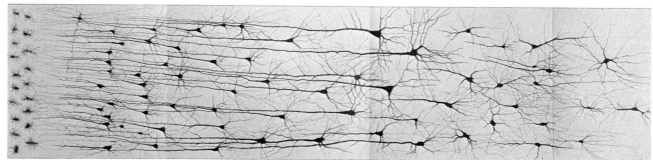

In the late 19th century an Italian physician, Camillo Golgi, developed a method of making brain cells clearly visible under the microscope. He used body *tissue*—a collection of similar cells which performs a single function. Staining a thin slice of brain tissue with a special preparation of silver salts, he found that this had the effect of picking out a selection of neurons complete with their intricate tracery of branches. More recently developed techniques of microscopy have been added to these beautiful Golgi-stained pictures of brain tissue to give us a fantastic view of the anatomy of the brain through greatly magnified pictures of individual cells. The modern *electron microscope* is capable of magnifications of up to half a million times, and by means of the scanning electron microscope we can produce pictures of the three-dimensional structure of the cell. An example of a picture taken by these methods, called a *photomicrograph*, is reproduced below. It is the brain cell of a rat.

In this photomicrograph the nucleus of the cell can be clearly

Exploring the Brain's Cells

Below: photomicrograph showing the brain cell of a rat. The nucleus is the central circle of coarsely granular appearance; the roughly triangular cell body fills the center and lower part of the picture.

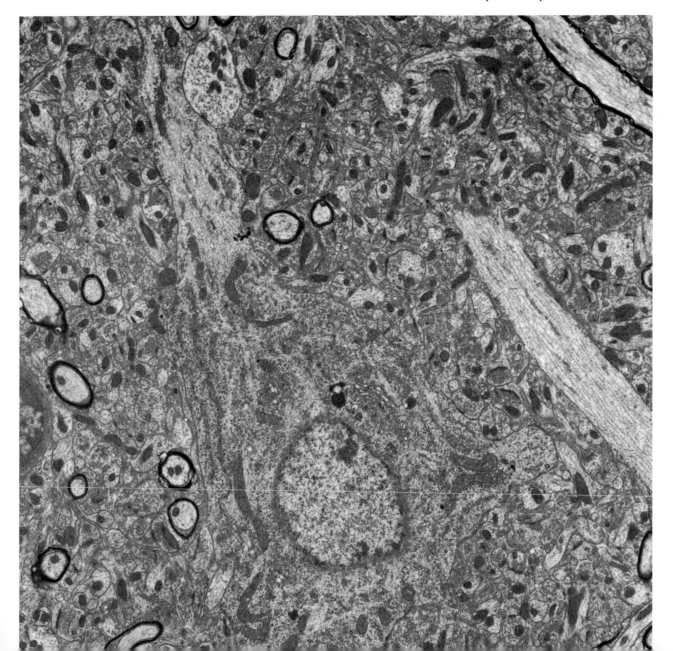

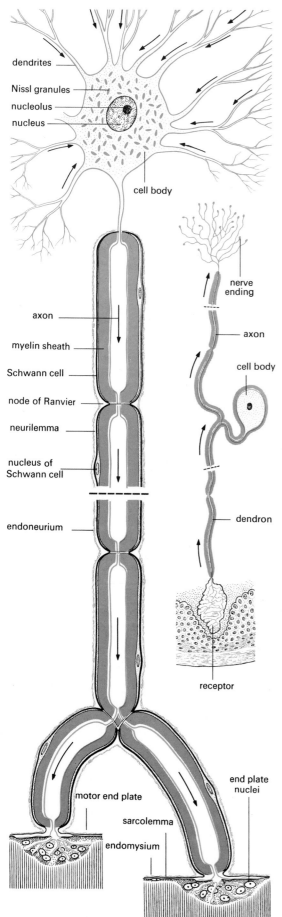

dendrites

Nissl granules

nucleolus

nucleus

cell body

nerve ending

axon

axon

myelin sheath

cell body

Schwann cell

node of Ranvier

neurilemma

dendron

nucleus of Schwann cell

endoneurium

receptor

end plate nuclei

motor end plate

sarcolemma

endomysium

seen. In the nucleus is the DNA, which codes the information for the protein production necessary for the cell to perform its functions. A nuclear *membrane*, a thin flexible covering, separates the nucleus from the main body of the cell, which is composed of *cytoplasm*. This is flecked with the *mitochondria*, which are the power packs of the cell, providing energy by processing *glucose*, a sugar. Grained all over the cytoplasm are the *ribosomes*, which contain the RNA, "messenger" of the DNA. In all these details the neuron is the same as any other body cell. However, in the photo some of its unique *dendrites*, or branches, are partially visible, by means of which the neuron interconnects with other neurons. The cell has been isolated from its surrounding material so that its basic structure may be seen. In order to visualize the entire brain we must think of billions upon billions of these cells, tightly layered and packed together.

In 1906 Camillo Golgi shared the Nobel Prize for Medicine or Physiology with a Spanish anatomist specializing in tissues, Santiago Ramón y Cajal. On the occasion of the presentation of the award, however, the two men would not speak to each other. The reason for their coolness was their conflicting views on the dendrites of the brain cell. Golgi believed that these innumerable branches connected every neuron in the brain to all the others, making the brain a tissue web of fantastic complexity. Ramón y Cajal, on the other hand, believed that each cell with its dendrites was separate from all the others, and that there was a small gap between them. The British physiologist Sir Charles Sherrington supported Cajal, and called the gaps *synapses*.

Cajal earned Golgi's unforgiving wrath by using the Italian's own staining technique to demonstrate his point. It was not until the electron microscope was developed in 1933, however, that the synaptic gaps became clearly visible and Cajal was proved right. The synapses, it is believed today, comprise about 10 percent of the brain's volume, and each neuron may have some 10,000 synaptic contacts with its neighbors.

The branches off the body of the cell constitute some 90 percent of the total cell volume in the brain. In addition to its dendrites, each neuron has a longer and thicker branch called the *axon*. Through the axon the brain receives its messages from and sends its instructions to the body's nervous system. Each axon is insulated with a fatty substance called myelin to protect it and to prevent the messages from getting confused. The axon in its protective sheath is called a nerve fiber. A single bundle of fibers, for instance that forming the optic nerve, may have as many as 10 million axons, each connected to a separate cell.

What happens during synaptic contact between cells is that a message is sent across the synaptic gap. This transmission depends on a complex electrochemical process, but can be compared to the spark jumping the gap in an automobile spark plug. A "transmitter substance" is dispatched from one cell to another via the dendrites. Although the synaptic gap is only about two millionths of a centimeter wide, the molecules of transmitter substance are even smaller. When they reach the other side of the gap they are absorbed by special "receptor molecules." When enough have been absorbed, an action is triggered in the "receptor cell," which is said to "fire." The "firing" of a bundle of

Axons, Neurons, and Dendrites

Left: Santiago Ramón y Cajal (1852–1934), the Spanish anatomist who with Camillo Golgi was joint winner of the 1906 Nobel Prize. One of his most fundamental achievements was establishing the neuron or nerve cell as the basic unit of the nervous system. Having adapted Golgi's silver-stain method of staining nervous tissue, he explored the then unknown areas of the cerebellum and cerebrum. This portrait accompanied an obituary notice written by his colleague, the British physiologist Sir Charles Sherrington.

Opposite: diagrams of a motor neuron (left) and a sensory neuron (right). Small projections called dendrites pass messages *into* the cell body while longer and thicker branches, the axons, convey impulses *from* the cell body out to motor end plates in the muscles. Nissl granules in the cell body contain RNA, probably for the synthesis of protein. The axon is surrounded by Schwann cells that secrete the fatty substance called myelin, which forms a protective sheath around the axon. Both of these layers are broken by frequent gaps called nodes of Ranvier, over which impulses must jump, thereby increasing their velocity. The outer membrane of the Schwann cells is called the neurilemma, and outside this is the fibrous endoneurium. An impulse travels in the opposite direction through the sensory neuron—towards the central nervous system. Starting from a receptor, the impulse passes along a dendron (really a long dendrite), through the cell body, and finally, by way of an axon, into the nerve endings of the central nervous system.

neurons leads to an instruction for potential action being sent down a nerve track to a particular part of the body. If an incoming message is being relayed, the "firing" results in the notification of a particular part of the brain of a situation reported by one of the body's sense receptors. The entire transmission process, of course, takes only a fraction of a second.

When we reflect that the brain neuron is just one of hundreds of different types of cells in the body whose form and function is coded in the DNA molecule, we can surely understand the joy and enthusiasm of Francis Crick. According to James Watson, as soon as they were certain that they had cracked the riddle of DNA, Crick "winged into the Eagle [a Cambridge pub] to tell everyone within hearing distance that we had found the secret of life."

44

EOPHAS FRATER CARNALIS IO=
PHI MARITI DIVAE VIRG MARIE

I
JACOBVS MINOR EPVS   MARIA CLEOPHÆ SOROR
HIEROSOLIMITANVS   VIRG MAR PVTATIVA MA
TERTERA D N

III   II
IOSEPH IVSTVS SIMON ZELOTES CONSO-
BRINVS DNI NRI

Chapter 3
Mistakes and Mutations

From "waltzing" mice to porcupine men—the incredible variation within a species is one of the most interesting qualities life can possess. What causes these "mistakes"—and are they always bad? What would have happened if we had all been born exactly like our parents from earliest times? Would the human race have lost anything important? Knowing how chromosomes work to map out future generations, will man ever be able to control such inherited diseases as cystic fibrosis or sickle-cell anemia? Almost all religions have placed taboos on marriages between close relations, and science confirms their good sense. But what of the modern dangers from radiation, and the equally modern attempts at genetic engineering?

"Waltzing mice" have been known as a curiosity for centuries. The animals' "waltzing" seizures are irregular and each lasts for several minutes. The mouse may revolve hundreds of times, turning to the left or right. If you or I, or indeed a normal mouse, attempted to do that, the result would probably be dizziness before the first 100 revolutions were completed. But the "waltzers" don't get dizzy. If they are spun in a machine they show no signs of dizziness, whereas normal mice are soon reeling all over the place. On the other hand, when placed in water they soon lose their sense of direction and drown. The "waltzers'" eccentric behavior has given a lot of people a lot of amusement over the centuries. However, it has only quite recently been realized that this behavior is due to a genetic deviation. "Waltzing" is hereditary.

Human dwarfs have given a lot of amusement, too. Every large circus has its dwarf clowns. Dwarfism, or *achondroplasia* as it is medically called, has been known throughout human history. Ancient statues of the Egyptian god Bes represent him with the stumpy legs and large head and eyes characteristic of dwarfism, and Bes was typically regarded as the buffoon among the gods. Dwarfs that survive infancy are near normal in physical fitness and intelligence, but few are fertile, so the genetic strain is not perpetuated by their breeding with each other. The majority of dwarfs are born to normal parents. It is evident that the gene causing dwarfism has been around for a long time and

Opposite: portrait of the Hapsburg Emperor Maximilian I of Austria and his family by Ambrogio de Predis. The Hapsburgs are a classic example of prognathism, a dominant inherited characteristic of excessive development of the lower jaw which causes it to project in front of the face. Pictured are members of three generations spanning a century from 1460 to 1560: Maximilian, his wife Mary of Burgundy, their son Philip the Handsome, and their grandchildren Charles V, Ferdinand I, and Mary of Hungary. The great-great-grandson of Charles V— Charles II of Spain (1661–1700)— manifested the "Hapsburg lip" in its most extreme form.

The Secrets of the Chromosome

Right: contemporary photograph of the famous British dwarf, General Tom Thumb, with his wife.

Above: Bes, the Egyptian god of recreation as represented by a statuette of the early XIXth Dynasty, around 1320–1280 B.C. He shows many of the physical attributes of dwarfism.

is very widely distributed among the human population.

A considerable number of human ailments and mutations are carried on the genes. Diversity is essential for human vigor and evolution, but the mechanism of haphazard genetic exchange that insures this diversity is also responsible for many of humanity's afflictions. It has been estimated that every human being is a carrier of on average three recessive mutant genes. But one of these will only become manifest and actually produce a child with a severe disability if it is paired with an identical recessive gene contributed by the second parent. The odds against this happening are long, and even when two people who have the same recessive gene marry, there is still an even chance that their child will be normal.

The cause of gene mutation is failure in the copying process of the chromosome prior to cell division. What causes such a failure to occur in nature is still something of a mystery, although it is known that it can be brought about by certain chemicals not normally found in the body, and by exposure of the sex glands to high doses of radiation. Sometimes an entire section of a chromosome, comprising many genes, undergoes a mutation. Over recent years considerable progress has been made in under-

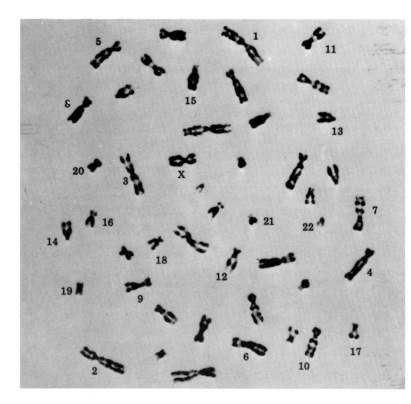

standing certain abnormalities in man in terms of chromosome mutations. Biochemists are now able to identify individual chromosomes and many of the characteristics associated with them. By isolating white blood cells, bone marrow cells, or skin cells, growing them in a laboratory culture and treating them with a chemical which slows down the process of cell division, they have been able to watch the chromosomes when they are spread out. By using a staining technique to bring out the bands that distinguish the pairs from each other, they have been able to classify them.

The male is distinguished from the female in humans in having 22 similar paired chromosomes and one pair which is conspicuously dissimilar, consisting of one long and one very short chromosome (designated X and Y respectively). In the female there is no difference between the 23rd pair, which is made up of two long chromosomes. The difference is expressed by saying that the male has the sex chromosome combination XY and the female has XX. A consequence of this is that the male partner determines a child's sex, for his sperm cell is composed of one chromosome from each of the 23 pairs, and the sex of his offspring depends on whether he transmits his X or his Y chromosome. If it is the X, he will produce a daughter; if the Y, a son. Normally half the male's sperm carry the X and the other half Y, so there is an even chance of a boy or a girl.

As a father always transmits his Y chromosome to his sons, certain genetic characteristics are sometimes passed on down the male line of a family while the females remain unaffected. An often-quoted example is that of the "Porcupine Men" produced in an English family in the 18th and early 19th centuries. The men had thick, black, very rough skin, in some places as thick

Above left: a montage of the 46 human chromosomes assembled in the laboratory so that each of the 22 similar pairs and the sexual or X factor (center) can be clearly seen.

Above: cells which have been irradiated with an acute level of X-ray exposure. The consequences of this treatment can be seen during reproduction in the upper four cells, in which the chromosome pairs have failed to move to opposite ends of the cell. This photomicrograph has been magnified approximately 1000 times.

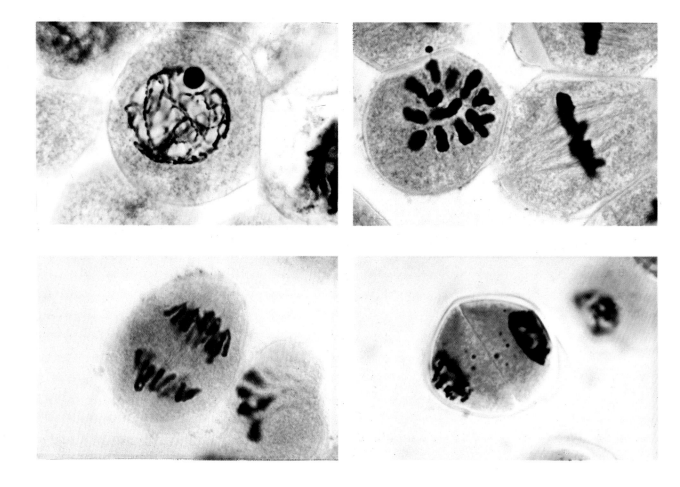

Above: photomicrographs of four different stages of normal cell meiosis as a part of sexual reproduction, when genetic information is exchanged and maternal and paternal characteristics are passed on to the offspring.

as half an inch. The first was born in 1716 in the village of Sapiston in Suffolk, and at least five males, possibly as many as 11, inherited the condition over the next four generations. This was a freak mutation—a more common genetic inheritance down the male line is a tendency to premature baldness.

It was once thought that the number of chromosomes characteristic of each species was fixed and unchanging, but recently it has been found that this law is not invariable. Some abnormalities are due to a person's having more or fewer chromosomes than normal. When an individual has more, the extra one is the same as those of an existing pair, and the person is said to be *trisomic* for that particular chromosome. When a person lacks one component of a pair, he is said to be *monosomic*. The irregularity occurs when a chromosome pair fails to separate in the process of cell division and either the two linked chromosomes enter the developing cell or neither of them does. Recent studies of miscarriages have established that in about 50 percent of cases, the fetuses have chromosome aberrations, an interesting example of how nature rejects its mistakes. When children are born with such abnormalities they often die in infancy, but if they do survive they grow up to show distinct abnormal characteristics relating to the chromosome affected.

Abnormalities of the sex chromosomes are the most common. For example, some men are born with distinct female characteristics. They have small testes, little facial or body hair, a feminine distribution of fat, and sometimes develop small breasts. Their condition is called Klinefelter's syndrome, after an American physician, Harry F. Klinefelter. Comparable characteristics are shown by a category of unfortunate females, who have under-developed genitalia and uterus and tendencies to dwarfism, deafness, and malformation of the heart. They are said to suffer from Turner's syndrome, which is named after an American gland specialist, Henry Herbert Turner. These conditions have long been known, but it was not until 1959 that their cause was understood. In that year Scottish biochemists John Strong and Patricia Jacobs in Edinburgh examined the cells of a man with Klinefelter's syndrome and found that he was trisomic for the sex chromosome. Instead of having the normal male combination XY, he had an extra female chromosome, making the combination XXY. This led to the examination of a Turner's syndrome sufferer, who was found to be monosomic for the sex chromosome—that is, to have just one X chromosome instead of the normal female combination XX.

These discoveries naturally raise the question whether any other abnormal combinations of the sex chromosomes exist. Several have since been found: females with the combinations XXX or XXXX, and males with XXXY, XYY, XXYY, and even XXXXY. The most common of these combinations is probably XYY, the extra chromosome conferring an extra element of "maleness," and the story of its discovery is interesting. A study was made of the inmates of a Scottish prison hospital for criminals difficult to control. Immediately nine men with the XYY combination were found, who were all exceptionally tall.

Chromosomes and Abnormalities

Left: a classic inherited characteristic. Baldness is a dominant trait in males but recessive in females.

Right: a rare inherited characteristic. Joseph Moerman, a Belgian, shows his six-fingered hands—he also has six toes on each foot. Polydactylism appears to be the result of a dominant gene, but although there appears to be no actual disadvantage involved, the condition is still very rare. It is common, however, in cats, which are often born with six or even seven toes.

Below: another inherited characteristic, seemingly not linked to survival—a white streak running through the hair at a particular spot.

Follow-up investigations in ordinary prisons established that a relatively high percentage of criminals over 6 feet tall were XYYs—at Nottingham, England, the figure was 20 percent, or one in five. The proportion in the normal population has been variously estimated at between one in 250 and one in 1000. The thought should be chastening to male chauvinistic pride that an extra Y chromosome apparently causes a tendency towards criminal behavior. This fact has been used by defence lawyers in murder trials, who have put in pleas of diminished responsibility because of their clients' chromosome abnormality. In 1969 a Los Angeles judge ruled that the XYY syndrome could not be used as a legal excuse for criminal acts.

Oddly enough, it seems that people who are tri-, multi-, or monosomic for the sex chromosome have a better chance of survival than those with deviations on other chromosomes. An extra chromosome of the type classified as number 21 is, unfortunately, not uncommon—about one in every 500 births has this abnormality. This condition is medically known as Down's syndrome after a 19th-century British doctor, Langdon Down, but is more commonly called mongolism. Its characteristics, which include a small, round head, small and slanting eye sockets, a small mouth with straight lips, stubby fingers, and general looseness of the joints, together with a tendency towards heart malformation and mental retardation, are of course not typical of the Mongolian race. The commonly used name for the affliction is an unfortunate survival of 19th-century insensitivity and prejudice, but Down's syndrome is a chromosome aberration which causes a great deal of distress. The discovery of its cause means that in future much of this distress may be avoided. It is now possible to determine whether a fetus has chromosome abnormalities early in a pregnancy by examining cells taken from the fluid of the uterus. If discovered at that stage, the woman can be offered the option of having the pregnancy terminated.

Characteristics attributable to recessive mutant genes are not so easy to detect before a child is born. "Gene maps" of the two sex chromosomes are gradually being filled in, but knowledge builds up slowly since human beings cannot be bred experimentally. There is also a considerable time lapse between generations, and sufficiently detailed clinical information about past generations is not always available. The only characteristic, apart from maleness, known to be associated with the Y chromosome is hairy ear rims. Rather more is known, however, about the genetic complement of the X chromosome, for it is considerably larger (see the picture on page 47) and therefore can carry more genes. This is why females are usually the carriers of recessive genes.

Records of royal and aristocratic families are valuable to geneticists for tracing the pattern of emergence of a specific gene over many generations. Royal family portraits of both male and female Hapsburgs, for instance, ranging from the 15th to the 19th century, show the same ugly characteristic of a protruding lower lip and narrow jaw. The "Hapsburg lip" is often put down to inbreeding, but this is not necessarily the case. Genetic deformities and diseases may be produced by mating between individuals with no family connections, and in fact often are. Royalty's genetic misfortunes are only better known because they have been better documented.

For instance, females of the European royal families have carried the gene for hemophilia, a defect that prevents blood from clotting, for the last 150 years, and male royalty has been plagued by the disease throughout the period. The eighth of Queen Victoria's nine children, Leopold, was a hemophiliac, and three of her daughters were carriers of the disease. So virtually half her offspring were affected, as might be expected, because a mother passes on one of her X chromosomes to each of her children. The chances of it being the defective one are 50–50 each time. Women who carry the defective gene will not normally suffer from the disease themselves, for they have another X chromosome which will supply the blood clotting mechanism. They will only be hemophiliacs if they inherit the defective gene from both parents, a remote possibility but one which has been known to occur. Queen Victoria's son Leopold survived to the age of 31, when a minor fall caused a fatal hemorrhage, but before that he fathered

Inherited Afflictions

Below: family tree showing the descendants of Queen Victoria who were afflicted with hemophilia. Four of Victoria's children carried the gene for hemophilia, and the son Leopold actually suffered from the disease, which makes it possible for its victims to bleed to death from even a small cut.

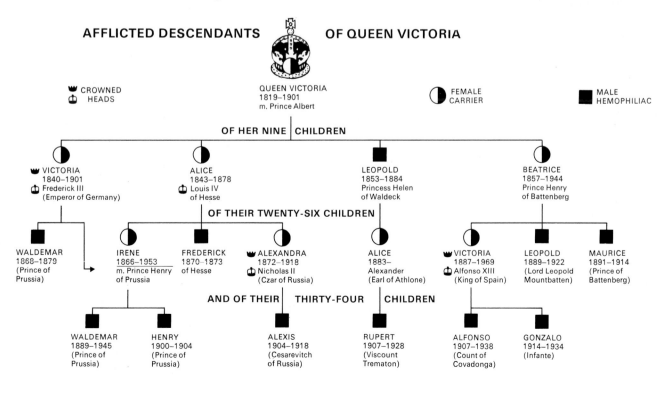

Hemophilia and Color Blindness

a son and daughter. His son inherited the affliction, and in due time produced a son of his own who died of the disease at the age of 21. Of Victoria's daughters, Alice produced one hemophiliac son and two carrier daughters, Beatrice two male victims and one female carrier, and the Queen's namesake Victoria bore two affected sons who died at the ages of two and 11. In the next generation Alice's carrier daughters between them bore three hemophiliac sons (including Alexis, the son of the last Russian tsar), and Beatrice's carrier daughter in her turn produced two hemophiliac sons and two daughters. The sons died childless of the disease, but the two daughters had children, and although their sons were fortunate enough not to inherit the defective gene, there is an even chance that the females are carriers and that hemophilia may reappear in the line in the future. The members of the present British royal family are unaffected, however, because they are descended from Victoria's son Edward, one of the fortunate ones who did not inherit the defective gene.

A much more common condition than hemophilia, passed on from generation to generation in the same way, is partial color blindness. Again, males are much more vulnerable to it than females, because it is carried on the X chromosome. The female's

Right: Queen Victoria (center) presiding over an 1894 family gathering in Coburg. Seventeen of the people in this group were her descendants, including Princess Alexandra of Hesse (standing behind and to the left of her grandmother the Queen), who would soon be married to Nicholas II, the last czar of Russia (standing beside her). Their son Alexis inherited hemophilia from his mother, who was a carrier.

Left: the Russian imperial family on their yacht. Alexis, heir to the throne, was afflicted with hemophilia from birth through his mother, who was a granddaughter of Queen Victoria. He lived his entire childhood in the shadow of the possibility of bleeding to death because of the inability of his blood to clot.

double X complement affords her a safeguard if she receives the defective gene from only one of her parents. It has been estimated that in Europe a full 7 percent of males and only 0.5 percent of the females suffer from partial color blindness. As the condition is not potentially fatal like hemophilia, the recessive gene responsible does not die out so easily. An interesting fact is that among more primitive peoples, such as Eskimos and Australian aborigines, the incidence of color blindness is much less, which suggests that the processes of natural selection are still at work suppressing those with defects who are less likely to cope successfully with life. As societies become more advanced and civilized, more genetic defects are likely to survive, both because natural selection ceases to operate so severely and because the afflicted receive better care.

In parts of Africa and India a genetically-transmitted disease called sickle-cell anemia afflicts about one child born in every 100. The disease affects the hemoglobin in the blood, the red pigment that carries iron and oxygen. It results in distortion of the red blood cells so that they become sickle-shaped. When the sick hemoglobin is analyzed, it is found to differ from healthy hemoglobin only in the substitution of one of its many amino acid components for another. A substitution of one base nucleotide for another in one of the numerous "word sequences" of three nucleotides in the DNA of the gene concerned is responsible. Even from such a small deviation a deadly disease can arise. The continuing high frequency of sickle-cell anemia, and its geographic distribution, have been explained by the fact that possession of the gene seems to provide a certain immunity to malaria. This disease continues to plague Africa and India, so children who receive the gene from only one parent (and are therefore not affected by it) have a survival advantage that normal children lack. It is surely one of nature's most ironical

Right: photomicrographs of red blood cells used in the study of sickle-cell anemia. Those containing normal hemoglobin (far right) are disk-shaped, while cells containing sickle-cell hemoglobin may become pointed.

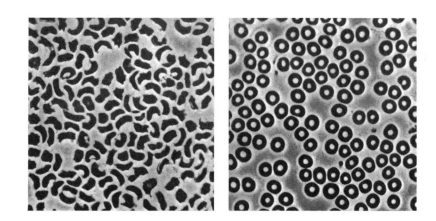

Below: a map showing the incidence of sickle-cell anemia. It is common in the areas of Africa, southern Europe, Arabia, and India which are shown in brown, but much rarer in the regions marked in gray.

injustices that the very people responsible for perpetuating one deadly disease should be rendered immune to another by virtue of the dormant killer gene they carry.

In Europe and North America the most common disease attributable to mutant genes is cystic fibrosis, a condition that affects about one birth in 1600. Sufferers develop blockages in their digestive and respiratory tracts which not only make digestion and breathing difficult but also give rise to infections. Until comparatively recently the disease was always fatal. The frequent occurrence of the disease is thought too high to be attributed to new mutations, and it has been suggested that carriers must acquire some advantage over non-carriers, as do the carriers of sickle-cell anemia. However, the nature of any such advantage has not yet been determined.

A rare genetic disease is Huntington's chorea, which is named after an American physician who in 1872 was the first to fully describe the disease. It involves deterioration of brain nerve tissue and consequently of mental processes, as well as loss of control of some body movements. Sufferers usually develop the condition in early middle age. Rare as it is, Huntington's chorea is well known, and it provides a good example of how a genetic peculiarity can spread in a population. Huntington was able to trace the history of the disease back to six immigrants to America from the village of Bures in Suffolk, England, who arrived in Boston, Massachusetts in 1630. A survey made in 1916 found a total of 962 cases in the United States, all recipients of the mutant gene passed down from the original Suffolk immigrants.

It is a common belief that aristocracy and the peasantry have in common a tendency to inbreed and consequently to produce offspring who are defective from birth. The fact that both the "Porcupine Men" and the first carriers of Huntington's chorea in America came from the eastern English county of Suffolk might support this view, for even to this day Suffolk is one of the most self-contained rural communities in England. Attitudes vary widely on the subject of the marriage and mating of relatives, but the genetic arguments against the practice in certain cases are clear and definite. As every human being carries recessive genes, and as members of the same family have a proportion of their genes in common, inbreeding greatly increases the chances of two recessives of the same kind being combined, thus producing offspring with the recessive characteristic. For in-

inbreeding are found in most cultures, ancient and modern, with only a few exceptions such as the rulers of ancient Egypt and the Incas of Peru. Although such taboos cannot have been based on the precise knowledge of genetic laws that we have today, they do show a wisdom probably born of observation. Prehistoric man may have noticed that marriages between close relatives were often ill-fated, and hence assumed that they were therefore disapproved of by the gods.

If close relatives breed and have the misfortune to combine the same recessive gene, their afflicted offspring will, sooner or later, carry that recessive gene to the grave. It will die out with them. So here is another of nature's ironies: if these types of matings

The Perils of Inbreeding

Below: ancient Egypt was one of the few societies which encouraged inbreeding among its rulers. Here the pharaoh Tutankhamun is shown with his sister Ankhesenamun, whom he married in order to inherit the throne. The carving is from the lid of a wooden chest inlaid with ivory dating from the XVIIIth Dynasty and comes from the tomb of Tutankhamun.

Radiation and Gene Mutation

diminish because of modern knowledge of the genetic dangers involved, fewer recessive genes will die out and instead the genes will be passed on unactivated but unhindered. The dangers will thereby increase for future generations. Marrying one's relatives is a practice almost universally frowned upon, and for good reason, but ironically it is in a sense a service to human posterity.

Right: early X-ray equipment—the orthodiagraph, which delineated the organs of the body and showed them in their natural size.

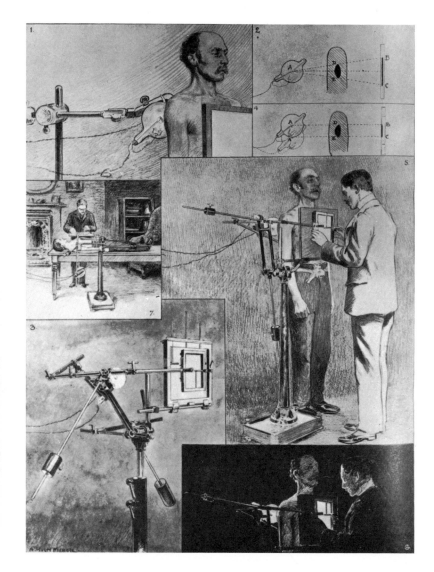

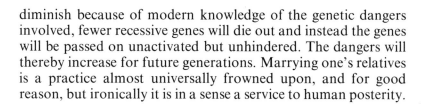

Above: the Polish scientist Marie Curie, winner of the Nobel Prize in 1911 for her discovery of polonium and radium. Although one of the first to work with radiation, she did not recognize its inherent dangers.

This is not a problem people are generally aware of, but the potential genetic danger to mankind that might be caused by radiation is a subject of common knowledge and concern. Radiation is the one definitely known cause of gene mutation, and over the last century the amount of radiation human beings are subject to has increased sharply. The increase began with the use of X-ray machines for medical purposes after the turn of the century. In the 1930s the British scientist J. B. S. Haldane calculated that the prevailing background high-energy radiation was the right amount to account for the rate of mutation at that time, and he warned that any increase would increase the mutation rate proportionately. The rate of increase was, however, greatly accelerated after the explosion of the first atomic bomb in 1945 and through the subsequent series of nuclear tests conducted by the major political powers until the end of 1962, when the Test Ban Treaty was signed. Scientists have not agreed as to the amount of radiation human beings can safely be exposed to, and as the question cannot be settled experimentally they will no doubt go on arguing about it. On the one hand it can be said that there is no such thing as a safe level, for any amount of radiation is potentially *mutagenic*—that is, capable of producing a mutation; on the other hand, we have to accept a certain level, both because radiation reaches us from space and from the earth itself, and because man-made radiation, including X-rays, is a source of many benefits. Man cannot afford to be complacent about the hazards of radiation, but at the same time he should not panic unnecessarily. It has been estimated that in the first 30 years of life a human being receives three units of radiation from natural sources and rather less than three units from man-made sources, and experiments with mice have established that it takes about 60 units, administered quickly, to double the mutation rate.

Just as radiation is at present used to eliminate cancerous tissue in the body, it is conceivable that in the future it might be focused on the DNA molecule to eliminate unwanted genes or to produce deliberate mutations. "Genetic engineering" has, in fact, already been used in agricultural technology. Occasionally a desirable mutation will be thrown up in the random process. If it is possible to cast aside the bad mutations and retain the good, one could improve or vary a breed of organism. The irradiation technique is useful in agriculture, but there are obvious objections to its use in human breeding. Such an application has been foreseen, however, by more than one futurist writer of fiction or demented would-be demagogue.

Eugenics, wrote Francis Galton, the 19th-century British anthropologist who invented the term, "is the science which deals with all the influences which improve the inborn qualities of a race; also with those that develop them to the utmost advantage . . . I conceive it to fall well within man's province to replace natural selection by other processes that are more merciful and not less effective. This is precisely the aim of eugenics." The "other processes" Galton had in mind were forms of selective breeding. What kind of incentives or penalties might be used to accomplish this was a question he did not attempt to suggest, but the idea was that somehow human beings who were richly endowed by nature should be paired and encouraged to breed, and the poorly endowed should be discouraged. Actually it wasn't a very original idea—the English ruling class had been practicing it for generations—but to give the practice a name, call it a science, and suggest its wider application was original and, some thought, dangerous. Aldous Huxley sounded the alarm in his

Right: exploding the atomic bomb at Bikini, an atoll in the western chain of the Marshall Islands in the central Pacific Ocean. Between 1946 and 1958 the United States used the tiny island as a site for the peaceful testing of atomic weapons; the 200 or so inhabitants were removed to nearby islands before the tests began. The top two inches of Bikini's topsoil were removed before replanting was allowed.

The Cecils of Hatfield House through five generations spanning the 16th to the 19th century. William Cecil (opposite, above left), chief minister of Queen Elizabeth I of England, founded the dynasty by marrying as his second wife Mildred Cooke (opposite, above right), a remarkably well-educated and well-connected woman. Their son Robert (opposite, below left), who became the first earl of Salisbury, maintained and even carried forward the Cecil influence. James Cecil (opposite, below right), the seventh earl and first marquess of Salisbury, helped to reinvigorate the family fortunes. He was respectably active in politics and married the flamboyant daughter of an Irish peer. Their son married Frances Mary Gascoyne (above left), cultured heiress to a great fortune, and added her name to his to become James Gascoyne-Cecil (above). Their second son Robert Cecil (left) was Prime Minister of Britain almost uninterruptedly for 17 years under Queen Victoria. The history of the Cecils could be considered an example of the revitalization of a basically vigorous stock through marriage.

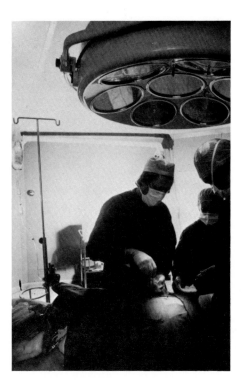

1932 novel *Brave New World,* and a gentleman who assumed political power in Germany the following year later continued the good work by putting forward a few of his own proposals, relating to Jews and gypsies, for using the science of eugenics.

Aiming to breed a superior race is known as "positive eugenics." The ethical problems raised by any proposal for the deliberate breeding of human beings are highly debatable. However, no such problems are raised by "negative eugenics," which would seek to diminish human distress and improve the conditions of life for many by eliminating genetic defects or their causes. Early identification of chromosome abnormalities in fetuses and termination of such pregnancies is one example of applied negative eugenics. Another example is marriage counseling based on knowledge of the genetic background of both prospective partners; such knowledge cannot be complete, but it can in some cases be enough to prevent potentially disastrous mismatchings.

The prevention of the conception or birth of children who would be defective because of genetic deviations does not pose great scientific or ethical problems, but when we consider tinkering with the DNA molecule the situation becomes more complex. Methods do not yet exist for performing such precise

Above: amniocentesis—the sampling of amniotic fluid from the womb of a pregnant woman—is now a standard technique for acquiring essential information about the health of the fetus. It consists of pushing a needle through the belly wall and into the womb to withdraw some of the fluid in which the fetus floats. A simple, low-risk procedure first suggested by a German named Henkel in 1919, here it is performed at a hospital in London.

Right: young girls give the Nazi salute at a rally in Germany in the early 1930s. Hitler idealized these "Aryan maidens" as part of his racist policy of positive eugenics.

microsurgery, and scientists are divided on whether such surgery ever will be possible. The 1962 British Nobel laureate Dr. Max Perutz of Cambridge has expressed the difficulties: "I fail to see how you are going to perform surgery on the genetic apparatus of man . . . The number of nucleotide base-pairs in a single human germ cell is in the order of 1000 million, distributed over 46 chromosomes. How could we delete a specific gene from a single chromosome, or add specific genes to it, or repair a mistake consisting of a single nucleotide pair in one gene? It seems hardly possible." It is only a bold scientist, or a rash one, who commits himself to a statement that something is impossible, and Perutz is cautious enough to add the qualification that "conceivably methods of transduction will become feasible; we may find harmless viruses which can be introduced into man and used to transduce desirable genes into people who lack them." Other scientists have been less cautious in their predictions. The 1958 American Nobel laureate, Professor Joshua Lederberg of the California Institute of Technology, has predicted that if scientific effort is focused on the problems, techniques of genetic engineering might be perfected in 10 or 20 years.

A beautiful woman dancer once said to the British dramatist George Bernard Shaw, "Think of a child with my body and your mind." Shaw, immune to flattery and always with his wits about him, replied, "Ah, but suppose it had my body and *your* mind?" He was right not to be enthused by the potential model of excellence the dancer promised, for up to 1950, when he died, breeding and genetic inheritance could not yet be controlled. But the situation is changing rapidly, and parents in the not very distant future may be able to choose not only the sex but also many of the characteristics of their children. Scientists are currently working hard at the problems, and in the meantime men and women everywhere should take the opportunity to consider the moral and emotional problems that these developments will undoubtedly bring.

Toward Genetic Engineering?

Below: Isadora Duncan, the beautiful and controversial modern dancer who died in 1927. She was intrigued by the possibilities of planned human breeding.

Left: caricature of the playwright George Bernard Shaw by Bernard Partridge for *Punch* magazine in 1925. Shaw was not impressed by Isadora Duncan's proposition.

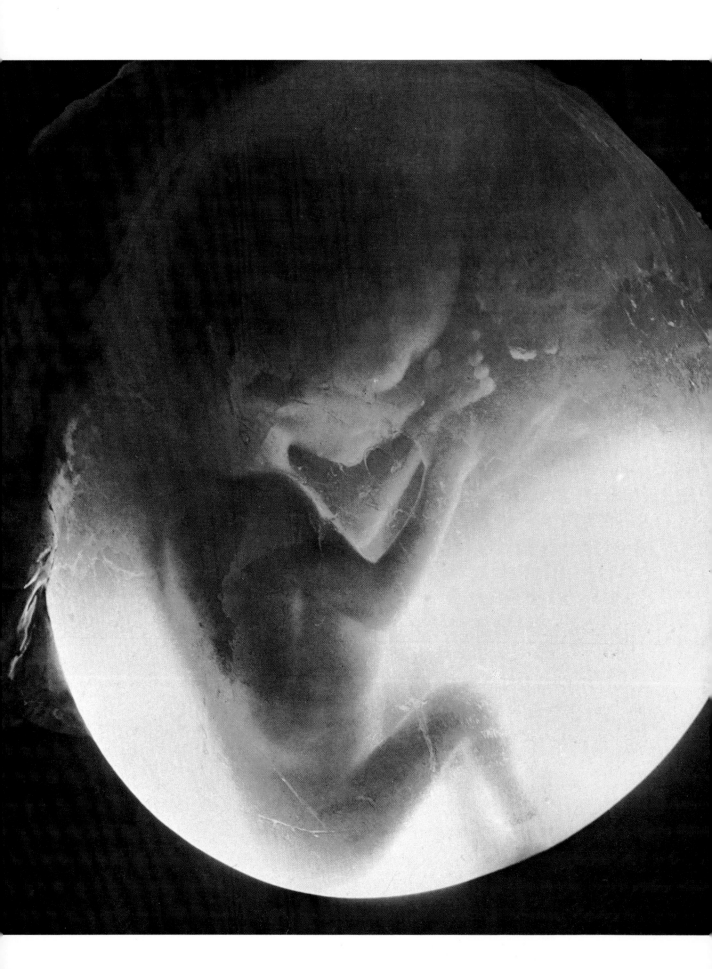

Chapter 4
The Developing Embryo

From the rare miracle of virgin birth to the common but no less miraculous process of ordinary sexual reproduction, we are constantly finding out more about the way new life is formed. The long journey of the egg, from production in the woman's ovary to a secure lodging in her uterus, is fraught with peril. But once the embryo begins to grow and slowly acquire human shape—is it safe from danger even then? When does its body become differentiated into face, lungs, feet, and toes, and why doesn't the mother's body reject the "foreign intruder"? In spite of man's age-old interest in his own pre-natal development, many mysteries still remain.

In 1955 a British mass-circulation newspaper appealed for any women who believed they had produced a virgin birth to volunteer themselves for scientific investigation. Nineteen volunteers came forward. Through a series of tests, 18 of them were eliminated. The one remaining was a Mrs. E. Jones, who had an 11-year-old daughter named Monica. Further tests on Mrs. Jones and the child to verify their relationship established that their blood, their saliva, and their tasting powers were identical. The only one of these tests that didn't produce a positive result was an attempt to graft skin from one to the other, but this didn't necessarily disprove Mrs. Jones' claim. After six months of tests, the newspaper announced that the results were consistent with a case of *parthenogenesis*, or virgin birth.

Parthenogenesis occurs in nature in lower biological organisms. Parthenogenic eggs, which have begun to develop by themselves without fertilization by male sperm, have been found in human ovaries, so the phenomenon cannot be dismissed as impossible, but on the other hand it is very difficult to prove. The most famous claim of its occurrence in history, that of Jesus of Nazareth, is particularly difficult to support scientifically, for virgin births should, according to the laws of genetics, result in daughters. The female's XX chromosome complement could only produce its like, although of course it can be argued that God in His unlimited power could have easily materialized the necessary Y chromosome. Be that as it may, generally speaking

Opposite: a human fetus. The miracle of reproduction has become less mysterious in recent years as medical technology has provided more information on what happens during the nine months of embryonic development.

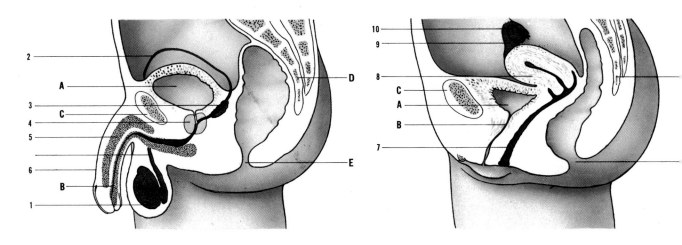

Above: the male and female reproductive systems. They possess certain attributes in common (A–E), but the main sexual organs (1–10) are complementary rather than similar.

A   Bladder
B   Urethra
C   Pubis
D   Spine
E   Anus

1   Testis
2   Vas deferens
3   Seminal vesicle
4   Prostate gland
5   Spongy tissue
6   Penis
7   Vagina
8   Uterus
9   Oviduct
10 Ovary

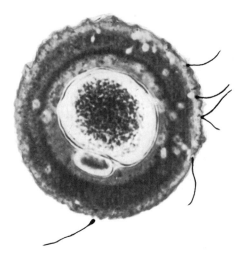

Above: a photomicrograph showing several sperms attempting to penetrate the human egg or ovum. The nucleus of the successful sperm fuses with that of the ovum to form the first cell of a new human being.

parthenogenesis is an unsatisfactory means of reproduction, not allowing the variation necessary for highly evolved and possibly still evolving mammals. Sexual reproduction is much more suitable. But considering the complexity of the sex process, it appears scarcely less miraculous than parthenogenesis.

One near miracle is that the male and female germ cells, the sperm and the egg, ever actually get together. The favorable time for their conjunction is very short. It must occur within a few hours of the female's egg leaving the ovary, at the very beginning of its 4-inch journey down the fallopian tube to the uterus (womb). The male's sperm must embark on a journey at the same time, a long and perilous journey which only one of the estimated 400 million sperm released in a single ejaculation stands a chance of completing. At the neck of the female's uterus is the protective cervix, which is usually covered with a thick mucus that prevents the sperm from penetrating from the vagina where the male deposits them. The sperm can only pass this barrier during a few hours of each month when the mucus becomes more liquid, but even then a majority fail to get through.

The thinning of the cervical mucus occurs at the point when the egg is in the upper part of the fallopian tube and ready to be fertilized. The sperm that have reached the uterus remain there for some time to undergo a process known as *capacitation*. What happens at this stage is not fully understood, but involves sperm contact with as yet unidentified chemicals. If the sperm do not become capacitated they will be unable to penetrate the outer layer of the egg if they reach it. After their sojourn in the uterus some of the sperm enter the fallopian tube on the last stage of the journey.

Covering the whole front portion of each sperm's head is a tiny cap known as the *acrosome*. This is a sac containing enzymes which are released when contact is made with the egg. These enzymes digest the material of the egg's protective outer layers, enabling the head of the sperm to embed itself and gradually force itself through. When one sperm has penetrated, a rapid change takes place in the egg's outer membrane which seals it off from all other comers. The successful sperm then penetrates deeper until it has breached the cell's wall, whereupon it releases its genetic material to pair with that of the egg. Fertilization has taken place.

The hazards to the conception of new life are by no means over by this stage. The fertilized egg must get itself into the nourishing environment of the womb and find a secure lodging there. But as it continues its journey down the fallopian tube it must divide and multiply itself several times, until it has become a small cluster of cells about the size of a pinpoint. Upon reaching the uterus, the cell cluster forms a shell, enclosing a tiny cavity filled with fluid, and at this stage the spherical structure becomes known as the *blastocyst*. About five days after fertilization, the blastocyst has reached the womb. What propels the fertilized egg on this journey down the fallopian tube is another mystery, and so is the fact that the journey time is about the same in all mammals, regardless of the great differences in distances traveled. For instance, the distance is 40 times longer for the embryo pig than for the embryo mouse.

The blastocyst, now approximately five days old, floats about the womb looking for a secure hold. This is a hazardous stage— it can, and sometimes does, float right out of the womb. This may happen if the wall of the womb has not been specially prepared to receive the blastocyst by the activity of hormones known as *estrogens*, which are produced in the ovaries. Its survival at this stage is also dependent on the production of another hormone by the ovaries, *progesterone*, which insures the continuance of its nourishment. The feeding layer of the egg—the *trophoblast*— sinks tiny grappling shoots known as *villi* into the wall of the womb. The developing organism has about a week to become implanted, because otherwise menstruation will occur and it might be swept away. In fact the blastocyst embeds itself so deeply that it completely disappears, and the woman's blood begins to flow through the outer layer of cells. The trophoblast will eventually become the *placenta*. By the time the woman's

The Miracle of Conception

Below left: the human uterus much enlarged by the presence of the young fetus inside. Surrounding the embryo are fetal membranes that enclose the amniotic fluid which protects the fetus.
Below: diagram of an ovum or female egg cell showing the outer membrane (a), the nucleus (b), and polar bodies (c). After fertilization the egg divides into two, then four, then eight cells, and so on. Later the differentiation of cells begins.

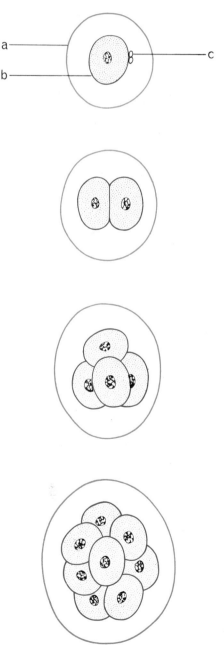

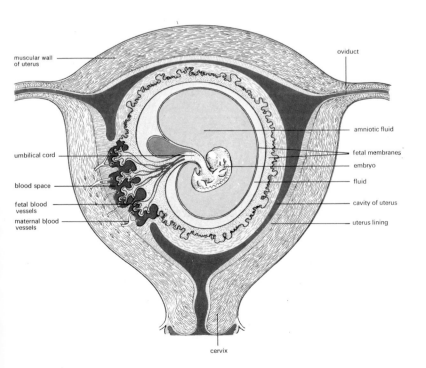

muscular wall of uterus

oviduct

umbilical cord

amniotic fluid

fetal membranes

embryo

blood space

fluid

fetal blood vessels

cavity of uterus

maternal blood vessels

uterus lining

cervix

The Mysteries of Pregnancy

Right: *The Immaculate Conception* by the 17th-century Spanish painter Diego Velazquez. Parthenogenesis or virgin birth has been considered a miracle and an omen throughout history. If it could indeed happen, the result should be a daughter, since the male Y sex chromosome would not exist among the mother's chromosome stock. (Reproduced by courtesy of the National Gallery, London.)

next menstruation normally would be due the embryo, measuring an eighth of an inch across, is safe. A special hormone circulating in the blood carries the message that prevents the triggering of menstruation. The many hazards to conception having been surmounted, the scene is set for an astonishing drama of explosive growth.

As one gynecologist has said, if a woman had a growth like that of pregnancy under any other circumstances, she would be considered seriously ill. It is one of the profound mysteries of pregnancy that the body can support such a rate of growth. Another mystery is that the woman's body tolerates the foreign body growing within it. For the new life is a foreign body; mother and child are two distinct people from the start. The placenta— the organ which develops to connect the embryo to the uterus— insures that their blood systems never mix, and in later life their systems will repel each other as vigorously as they would repel any other foreign body. A fetus is quite literally a graft onto the mother, and science is unable to explain why she doesn't reject it. In a discussion of this question, the British zoologist Sir Peter Medawar has put forward a number of possible explanations and then shown that each of them is inapplicable. He comes to the logical conclusion that, in the light of all our present knowledge, pregnancy is impossible.

But it happens; and how it happens is one of the marvels of life, which we have only comparatively recently been able to appreciate. Although man has always been fascinated by the question of his own origins, until techniques of microscopy were developed he was powerless to do anything but speculate about it. Leeuwenhoek, a Dutchman, first observed human sperm under the microscope in the 17th century, and for a long time thereafter biologists were divided into two schools: the "spermists" who believed that each sperm contained a *homunculus* ("little man"), a complete human being in miniature which the woman's body sheltered and nourished while it grew, and the "ovists" who believed that a homunculus existed in the female's egg and the sperm merely "awakened" it and incited its expansion to full size.

It was not until 1930 that the first human egg cell was seen coming from the ovary. In 1944 further progress was made when the union of sperm and egg was observed for the first time, and in 1952 the first blastocyst was seen. Knowledge of the drama, dangers, and wonders of the first weeks of life is thus very recent. A further breakthrough came in the 1960s when techniques of intrauterine photography—taking pictures inside the womb— were pioneered by the Swedish medical photographer Lennart Nilsson. Nilsson's fantastic and beautiful pictures of life before birth gave the people most concerned about pregnancy—pregnant women themselves, and their husbands—an unprecedented opportunity to know and visualize how the life they had created was day by day and week by week developing into a unique human being.

The actual creation of a human being takes only about eight weeks from fertilization. By the end of that period the cells of the implanted embryo have differentiated to form a *fetus*, a recognizable human being with all its organs, bone structures, and tissues complete but incredibly miniature. The entire creature is no bigger than the first two joints of an average adult's index finger.

Every college biology student learns that "ontogeny recapitulates phylogeny." The phrase was coined by the German biologist Ernst Haeckel in 1866, and means that in the development of the individual organism all the stages the species went through in its evolution are repeated. In other words, the embryo resembles in turn a fish, an amphibian, a reptile, and a primitive mammal before becoming a human being. This is true to an extent, but what happens in fact is that in the process of human cell differentiation the characteristics common to different species stand out more prominently at different stages of development.

Soon after the blastocyst has become embedded in the wall of the womb, three distinct layers of cells begin to develop on one side of it. These are known as the *ectoderm*, the *mesoderm*, and the *endoderm* (or *entoderm*). The outer layer, the ectoderm, builds up rapidly through cell growth and begins to fold over to enclose the other two layers in a kind of tube. Four weeks after conception this curved tube certainly looks nothing like a human

Below: *"del pissciare del putto"* begins his description of this sketch by Leonardo da Vinci. Although he once examined an actual fetus, he still assumed, mistakenly, that the infant was immobile, with one heel shutting off the release of urine before birth.

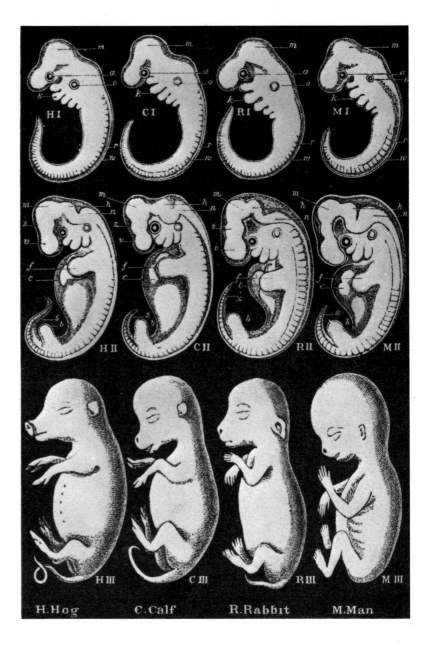

Right: Ernst Haeckel, the German biologist who coined the phrase "ontogeny recapitulates phylogeny," was an enthusiastic supporter of Darwin's theory of evolution. He compared the embryos of four different mammals (pig, ox, rabbit, and man) at three different stages of evolution and pointed out their striking similarities.

being. In fact it is almost identical to the embryo of the sea urchin or that of the chicken. It becomes more fishlike with the development of six arched projections on the ectoderm which look like gills. Above these arches a blunt head has begun to form, with spaces for the eyes, ears, and mouth becoming differentiated. The arches will eventually become not gills but the upper and lower jaws and throat of the human being. At four weeks the embryo is about a quarter of an inch long, but in those weeks it has grown at a tremendous rate to become 10,000 times bigger than the egg.

The outer layer of cells comprising the ectoderm will differentiate into skin, hair, nails, teeth, and all the features of the outside of the body. But the ectoderm also forms another tube, running from head to tail and swelling at the head. This is the embryonic spinal cord, nervous system, and brain. On each side of this inner neural tube blocks of mesoderm, the middle layer of cells, develop in pairs. These are called *somites* and will form the skeleton. The mesoderm also forms the muscles, blood, and some of the internal organs, including the kidneys and sex glands. The inner layer of cells, the endoderm or entoderm, forms most of the digestive system, such as the alimentary canal, stomach, intestine, bladder, lining of the lungs, and certain other glands, such as the thyroid and thymus.

By the 25th day after conception, when the mother is beginning seriously to wonder whether she is pregnant, a heart and system of blood circulation have formed. They are rudimentary at first, but the blood vessels spread out rapidly, for the entire embryo is no bigger than a pea, although the heart is large in proportion. By the fifth week, the head has lifted from its bent-over position, looking as if it had its heart in its mouth, and some resemblances to human shape and features have begun to appear. The tail is falling behind in its development, while little swelling buds have appeared on the main body and have begun taking shape as arms, hands, and feet. The cells of the ectoderm, mesoderm, and endoderm are all multiplying at a rapid rate and differentiating in form and function in obedience to the instructions coded in their DNA.

In the sixth week of life facial features begin to form. At first the embryo's eyes look out to the side like those of a fish, but as the head is modeled behind and around them they move to a frontal position. The tip of the nose becomes visible and eyelids begin to form. At the same time there is substantial skeletal development, and the separate fingers and toes can be distinguished. The embryo is now growing at a rate of a millimeter a day. This rate of growth continues through the seventh week, when liver and kidneys begin to work and further rapid development of the head takes place, the upper and lower jaws becoming visible and the ears, lips, tongue, and first buds of teeth forming. By the end of one more week, the eighth, the process of embryonic development is virtually complete and the tiny creature has become a fetus, a whole human being barely an inch long with but one purpose: to grow and grow and grow.

In the ninth week spontaneous movements begin to occur,

Amazing Growth

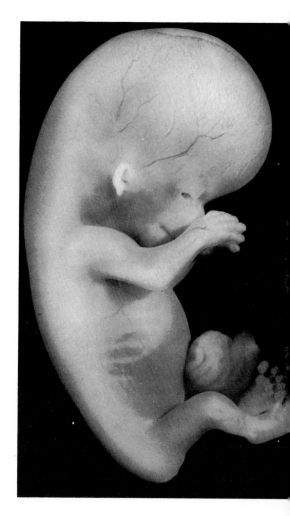

Above: the human fetus at 8½ weeks after ovulation. By the end of the eighth week the tiny creature is virtually complete, although barely an inch long. Spontaneous movements begin and the sex of the child-to-be can be determined.

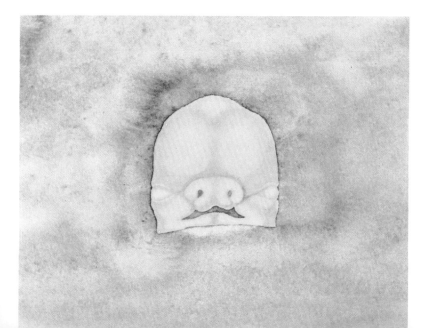

Left: the human head by the seventh week after conception. The neck has begun to take shape; the eyes, lidless like a bird's, are located on either side of the head. The nose is flat, but the nostrils are moving closer together. The mouth is still very broad, extending across the lower part of the face. The ears are still very small and are located far down on the sides of the head.

A Tight Schedule of Development

Right: ultrasonic scanning of the womb, a modern technique for monitoring the development of the unborn child. The machine produces an image of the fetus which can be examined without discomfort to the mother, from five weeks through to term. The mother's body is scanned by a very high-frequency sound wave, which is partially reflected when it reaches the boundary between different types of tissue.

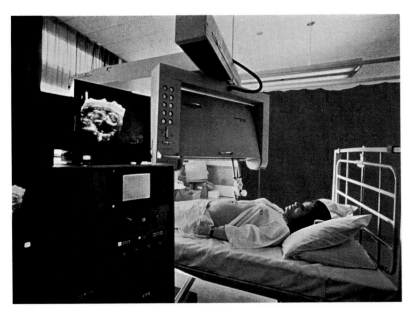

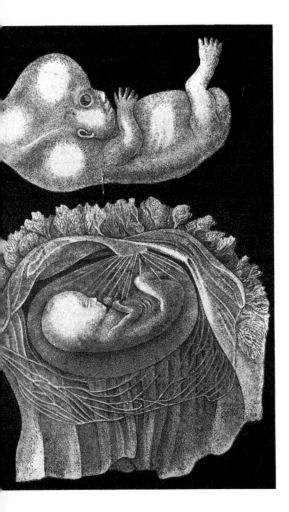

Above: representation of two human embryos, one of nine weeks, the other of twelve. The illustration comes from Ernst Haeckel's *Evolution of Man*, published in 1879.

although the mother will not yet feel them, and the sex of the child-to-be can be determined. The eyelids close over the eyes for the first time and the little hands start to make reflex gripping movements. By the 66th day the first quarter of the pregnancy period has gone. In the remaining three quarters the fetus will have to multiply its weight some 600 times. The mother will normally now be aware of her condition, although her weight gain because of the pregnancy is not usually more than one pound at this stage.

The tenth week after conception, the time of the woman's second missed menstruation, is when most miscarriages occur. When a fetus aborts naturally, it is usually because something has gone wrong in the process of growth—chromosomal abnormalities are a frequent cause of miscarriage. Malformations may also be due to the interruption of growth at some point. The schedule of embryonic development during those first eight weeks is tight and specific; if the development of any part falls behind schedule the loss cannot be made up and that part will either be missing or defective. In 1962 it was discovered that the drug thalidomide, which thousands of women had taken to counteract early pregnancy nausea, had the effect of interrupting the building schedule. However, the discovery came too late to prevent some 7000 live births of deformed children, most of them in West Germany.

It is an amazing fact that most children are born sound and healthy, considering the number of things that can go wrong at the embryo stage. In Britain in 1971, for example, congenital malformations were reported in only 1.7 percent of all births, both live and still. *Congenital* means at birth, as distinct from *genetical*, which means inherited; malformations due to genetic anomalies are included in the figure but would have been responsible for only a relatively small proportion of it. It is known that if a mother has German measles in early pregnancy it can affect the building schedule, but no other major causes of congenital malformation are at present understood.

The most common categories of defect are those of the nervous and circulatory system. Sometimes the curved tube formed by the ectoderm fails to close properly, and a child may be born with the condition known as spina bifida. There is a small gap in the baby's back, covered only by a membrane, through which cerebro-spinal fluid may be leaking. The gap can now be surgically closed, but the condition also affects the nerves that control the legs and bladder, so a child who survives may be incontinent and have difficulty walking.

Hydrocephalus, or what is commonly called water on the brain, is often associated with spina bifida. Excessive liquid accumulates in the cranial cavity because of a blockage in the circulation of cerebro-spinal fluid, and the child's head grows larger and larger. If the condition develops before birth, the mother will have a difficult delivery. Occurring after birth, it is often fatal, although there is now a valve that can be placed in the cranial cavity to draw the excess fluid away. The valve was invented by an American engineer whose child was a victim, and it has saved many lives.

The human brain is formed when the inner ectodermal tube balloons at one end to form three distinct swellings, which will become the forebrain, midbrain, and hindbrain. But sometimes this ballooning fails to occur, in which case a child with the condition known as anencephaly may be born. Completely lacking brain tissue, such births never survive. But if only one or two of the brain regions fails to form or if they are stunted, the child might survive, a victim of the condition known as microcephaly.

Malformations of the heart are the most common congenital defects after those of the nervous system, and are by far the largest category of defects of the circulatory system. The heart is a four-chambered pump in which each chamber is separate and designed to enable the blood to circulate in a particular way. When a baby is born its oxygen supply immediately starts coming from its lungs instead of through the umbilical cord from the mother's bloodstream, which means that the direction of the blood flow is suddenly changed. This requires a complex arrangement of heart tissues, and in their formation mistakes can easily happen. Sometimes the walls separating the chambers of the heart have gaps in them, in which case the child has a "hole in the heart" condition, and a serious case can produce a "blue baby," that is, one whose blood does not get enough oxygen.

Anomalies in the growth of the inner layer of tissue in the embryo, the endoderm, can cause malformations which only become obvious after birth when the baby tries to feed. The intestines may be entwined around each other or otherwise obstructed, the wall separating the windpipe from the gullet may be missing, or the anus may be blocked. These conditions can often be rectified surgically in the first days after birth. Surgery can also help other congenital defectives such as sufferers from cleft palate, hare lip, club foot, or dislocation of the hip.

After this period of high likelihood of miscarriage, between the 11th and 14th weeks, the fetus' body begins to resemble that of an infant. Its heart now pumps 50 pints of blood a day. At the end of the 14th week it measures 2.75 inches and weighs three-quarters of an ounce. Between the 15th and 18th week hair starts

Below: the human embryo, 13 weeks after ovulation.

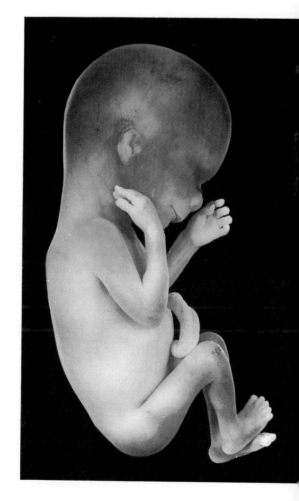

to grow, nails harden, nipples appear, and the mother can begin to feel movements. With the aid of a stethoscope that most obvious sign of new life, the heartbeat, can be heard. In this three-week period the fetus grows to a weight of 11 ounces and a length of 9 inches.

In the next three-week period, the 19th to the 22nd, weight increases to 1 lb 6 ounces and length to 12 inches. Fat begins to fill out the body and face contours, and the baby can grip with its hands. It also develops a curious growth of very fine body hair, known as *lanugo*, which vanishes before birth but meanwhile makes the fetus look like a furry mammal. Between the 23rd and 26th weeks weight increases to 2 lb 11 ounces, and the fetus may start thumb-sucking. Between the 27th and 30th weeks it generally settles into a quiet period, in a head-down position, and increases in weight to about 3 lb 11 ounces, nearly half the average final weight with eight to ten weeks remaining. There is a good chance of survival for premature births at this stage in their development.

In the last weeks of pregnancy the heart is pumping up to 600 pints of blood a day, although the baby only has just over half a pint. The fetus grows to a length of about 20 inches and is now too large to move very much in its confined space, although it may kick fairly vigorously. The growth process stops entirely a short time before birth as mother and child prepare to undergo the final drama and danger of delivery.

One of the unsolved mysteries of life is what actually triggers birth. A change in hormone secretions appears to have something to do with it. Just before labor starts, the level of estrogen tends

Below: the first contractions of the uterus signal the start of labor. Thus begins the 4-inch journey down the birth canal, ending in the moment of birth.

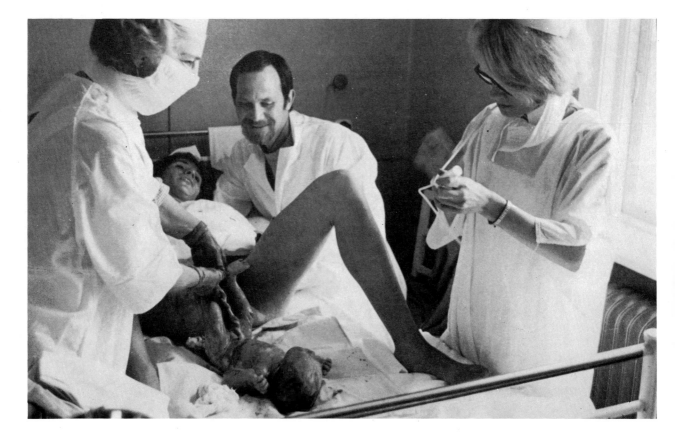

to drop and that of the hormone oxytocin rises—oxytocin injection is one way of inducing labor when it is overdue. Contractions of the uterus, occurring with increasing frequency, announce unmistakably to the mother that her labor has started, and thus begins what has been called the most dangerous journey in the world, the 4-inch journey down the birth canal.

However, it is infinitely less dangerous than it used to be. In 1900 the infant mortality rate in Britain was 154 per 1000, and in 1971 it was 17.5 per 1000. Maternal deaths in childbirth in the United States in 1915 were 6 per 1000, and in 1960 they were 0.25 per 1000. So modern mothers can look forward to the experience of childbearing with little fear. Narrow-minded taboos on the subject having been largely swept away, many mothers today can, despite the pain they inevitably undergo, find the experience emotionally rewarding, a profound personal experience of participation in one of the most common but enduring mysteries of life.

Birth: The Most Perilous Journey

Left: a mother and child. The experience of childbirth can be an emotionally rewarding one, the beginning of a lifelong attachment between two people who shared one body for nine months.

Chapter 5
A Pattern for Survival

The lifespan of a man has been compared to the swift flight of a sparrow from one area of darkness to another. The darkness of man's origins has occupied his attention as much as has his future: where did we come from, how did we get here, and why? From creation myths to evolution theory, answers have been offered. Darwin's revolutionary theory called forth a storm of protest from those who refused to accept their relationship to the apes, and anthropologists have been searching ever since for evidence of the "missing link," fossils of the half-ape, half-man creature who must have existed nearly a million years ago. Have they found him?

The 8th-century British historian, The Venerable Bede, records how a certain English king sought his councillors' opinions of Christianity. One of them replied with a speech comparing the life of man on earth to the flight of a sparrow through a warm and lighted hall in midwinter—making a swift passage from darkness to darkness. "Thus the life of man is visible for a moment," the councillor concluded, "but we know not what comes before it or what follows it. If, then, this new doctrine brings something more of certainty, it deserves to be followed."

Resigned to the mystery of man's past and future, Western man embraced Christianity not because it could be proved true but because it provided comforting answers. With the publication of *On the Origin of Species* by the British naturalist Charles Darwin in 1859, not only was a breakthrough announced which was to be as influential as Watson and Crick's DNA discovery a century later. The ideas involved also meant rejection of many beliefs which had prevailed up to that time. Until then, as the contemporary American anthropologist Loren Eiseley has said, "Man was a creature without history, and for a thinking being to be without history is to make him a fabricator of illusions." Darwin's theories shattered many of man's most comforting and flattering illusions, but they compensated for the loss by illuminating for him part of the darkness that had so oppressed that Saxon councillor's mind concerning the knowledge of man's history and evolution on earth.

Opposite: reconstructing the dodo. The dodo became extinct about 1680, but some skeletons remained. This painting shows a specimen being reconstructed at the Museum of Natural History in Paris. Interest in zoological classification and study helped to form the atmosphere in which Darwin composed his theory of natural selection; the extinction of a species showed the process of evolution at work.

The Great Debate

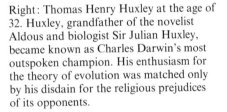

Right: Thomas Henry Huxley at the age of 32. Huxley, grandfather of the novelist Aldous and biologist Sir Julian Huxley, became known as Charles Darwin's most outspoken champion. His enthusiasm for the theory of evolution was matched only by his disdain for the religious prejudices of its opponents.

Darwin must have been the most timid man ever to bring about a revolution in human thought. Not for him the combativeness of Thomas Henry Huxley, the British philosopher and propagandist who became known as "Darwin's Bulldog." Huxley once said to his wife before giving a public lecture, "By next Friday evening they will all be convinced that they are monkeys!" So timid, in fact, was Darwin that he nursed his theories like a guilty secret for 20 years before publishing, and even then he tried to lessen the impact by omitting all mention of man except in his concluding noncommittal sentence: "Light will be thrown on the origin of man and his history." But his caution was to no avail. With his *Origin of Species* Darwin started a revolution which, like all revolutions, attracted advocates prepared to argue forcefully for the new ideas. When Huxley confronted Bishop Samuel Wilberforce in debate, the Bishop sneeringly inquired whether his ape ancestor was on his grandfather's or his grandmother's side. Huxley replied in words to the effect that he would rather have an ape for a grandfather than a pious hypocrite, and Darwin protested mildly that the contestants were "growling at each other in a manner which is far from gentlemanlike." The understatement was typical of him, just as in the *Origin* he had understated the case for the evolution of man as opposed to other species. But when the world is ripe for a revolutionary idea, it has a momentum that no one man's scruples or inhibitions can halt. Evolution was such an idea.

The idea of evolution had actually been around for decades before Darwin cemented all the theories and evidence with such

Left: a contemporary caricature, from the
London Sketch Book, of Charles Darwin
embracing an ape "relative." The ridicule
Darwin and his co-evolutionists received
was often scathing and merciless—it was the
reaction of people who felt that the
foundations of their self-image were being
threatened.

scrupulous scholarship that it could not be refuted. But before his time, the problems it posed for Christian believers were too great to be overcome. In the Bible, Genesis II said that man was God's supreme creation and that the animals were created after him for his companionship. This doctrine was irreconcilable with the evolutionary claim that man was himself an animal— a superior primate—and had appeared on the scene later. Furthermore, an all-powerful and all-seeing God would surely not have needed as roundabout a way to form his favorite creature as the evolutionists claimed had happened. Then there was the problem of time. For the evolutionary process to have taken place, the earth would have had to be millions of years old at least. In the 17th century Archbishop Ussher of Armagh in Northern Ireland had clearly shown that the world had been created in 4004 B.C. He based his calculations on the ages of the

The Christian Standpoint

Left: *God Creating the Animals* by the 16th-century Flemish painter Pieter Breughel the Elder. This painting, like many of its time and later, expresses the orthodox Christian view of the creation. God was personally responsible for each species, and afterwards He created Adam and Eve, who stand in the background amidst the creatures.

generations that succeeded Adam, according to the Old Testament record. Moreover, in the six days of creation God was believed to have made each creature perfect in every detail and to have ordained its eternal place in a "Scale of Nature." To believe that a species could change and that in the course of history some had even become extinct was virtually a blasphemy. It should have been a blasphemy, too, to believe in the existence of marsupials, an order of animals found only in Australasia, because the scriptures claimed that the world's animal forms had all moved outward from the mountains of Ararat in eastern Turkey, after Noah's ark had been grounded there. But the existence of the marsupials was a matter not of belief but of observation, so this problem had to be disposed of by pious reference to the inscrutability of God's ways. As objective scientific observation continued to turn up facts that were incompatible

"Evolutionists" Before Darwin

with scripture, devout Christians became increasingly pious, stubborn, and abusive, while some scientists, such as Darwin, were embarrassed and hesitant, and others, such as Huxley, grew exultant.

In 1726 a Zurich doctor, Johann Scheuchzer, had discovered a fossil skeleton which he promptly named "Diluvial Man" and described as "the bony skeleton of one of those infamous men whose sins brought upon the world the dire misfortune of the Deluge." Scheuchzer asserted that the Flood had occurred in 2306 B.C., and with the assistance of enthusiastic followers he made a collection of fossils for a "Museum Diluvianum," an exhibition of supposed relics of that biblical event. Comical as Scheuchzer's endeavor may seem, it was actually a forward step in his day, for he claimed that fossils were not haphazard "sports of nature," as had been believed, but the petrified remains of formerly living things.

The most famous natural scientists of the 18th century were the Swede Carolus Linnaeus and the Frenchman the Comte de Buffon. Linnaeus painstakingly classified all the phenomena of the natural world, considering his work a celebration of the

Right: Johann Scheuchzer, the Zurich doctor who discovered a fossil skeleton in 1726 which he labeled "Diluvial Man." He tried to attach an exact date to the biblical Flood and was very influential in his day.

Opposite (left): Georges Louis Leclerc, Comte de Buffon (1707–88), the French naturalist who wrote a comprehensive work on natural history. His *Histoire Naturelle* was the first to present apparently disconnected facts in an understandable form.

bounty of God as expressed in His creation. Linnaeus believed the inhabitants of the natural world had remained unchanged since those momentous six days. He dismissed fossils as mere stones and did not attempt to explain their origins, but, interestingly, he classified man in the order of primates, together with the great apes. No idea of evolution was meant by this classification, though, for it was a rule of his system that the lines of demarcation between the species were fixed and unchanging.

The Comte de Buffon was Linnaeus' greatest contemporary. Although he couldn't match the Swede as a classifier, he had a Gallic inclination for theorizing, and scattered throughout his works are quite a few ideas that look forward to the next century. Buffon rejected biblical chronology and declared that the earth must be at least 75,000 years old, perhaps very much older. He observed the struggle for existence between living things, but did not make the Darwinian deduction that the struggle played a part in any sort of natural selection. He recognized that there were variations within a species and that by deliberate selection and breeding desired characteristics could be produced. He even made the blasphemous statement that "In order to improve

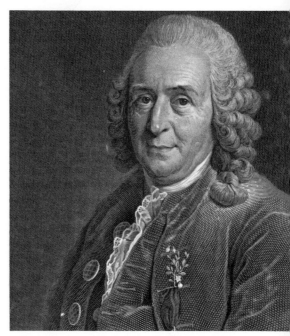

Above: Carolus Linnaeus (1707–78), Swedish botanist known for his comprehensive systems of classification for all the observable phenomena of the natural world. Although not convinced that species evolve and therefore unimpressed by the evidence of fossil finds, his classification systems helped form the basis of modern biology.

Above: illustration from Volume VII of the Comte de Buffon's *Histoire Naturelle* (1753), showing an orang-utan. Buffon noted the similarities between men and apes a century before Darwin.

Above: Baron Georges Cuvier (1769–1832), French student of Buffon and founder of palaeontology. In 1796 he read a paper "On the species of living and fossil elephants" in Paris, in which he presented for the first time detailed and almost irrefutable evidence for the concept of extinction. He also built a remarkable reconstruction in 1806 of the American fossil mastodon.

Nature, we must advance by gradual steps." Buffon also observed the basic similarities between mammals, and he asked: "At what distance from man shall we place the large apes, who resemble him so perfectly in conformation of body?" He paid dutiful Christian homage to a divine spirit in man which distinguished him from all other creatures, but his scientific observations continually led him to make heretical speculations, and in fact he produced, a century before Darwin, many of the basic ideas of the Darwinian theory.

Two of Buffon's pupils, the Frenchmen Baron George Cuvier and Jean Baptiste Lamarck, also made notable contributions to the study and theory of human origins in the 19th century. Cuvier became known as the founder of *palaeontology*, the science dealing with fossil remains—his skill in reconstructing entire prehistoric skeletons from fragments earned him the title "magician of the charnel house." He examined the bones of Scheuchzer's "Diluvial Man" and pronounced them the remains of a giant salamander. His studies revealed that there was a time, not very remote geologically speaking, when man had not existed on the face of the earth, though many other forms of life had. He also observed that the creation was not a ladder leading up to man but a great tree with branches that had developed independently. Although later discoveries in palaeontology were to prove wrong his statement that "There is no such thing as fossil man" and to put back the date of man's appearance on earth far before his estimates, and although Cuvier himself was opposed to evolution theory, the science he founded was to make great contributions to the defense of the evolutionary hypothesis.

Lamarck was a convinced evolutionist who avoided some of the theological problems of the theory by claiming evolution was "the accomplishment of an immanent purpose to perfect the

Right: illustration of Baron Cuvier's reconstruction of the hippopotamus, published in 1824. It was Cuvier's skill as an anatomist that made him such a seminal figure in palaeontology—he was reportedly able to reconstruct an entire animal from one small fragment of bone.

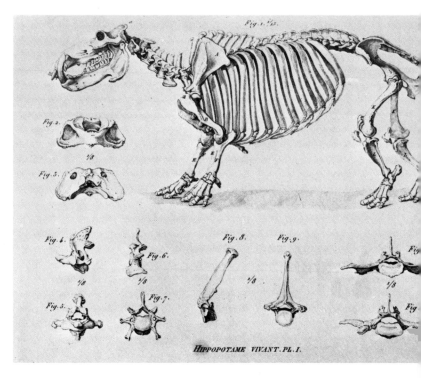

HIPPOPOTAME VIVANT. PL. I.

creation." He saw life as having an inherent, God-given drive towards perfection by way of increasing complexity—an urge to ascend through higher and higher levels. The idea was appealing to an age that sought a scientific basis for religion, and later in the 19th century the Irish dramatist and critic George Bernard Shaw was to declare himself a Lamarckian and develop an influential philosophy-religion around his idea. Its appeal was the introduction of the concepts of will and striving into evolutionary theory. Lamarck believed that, over the course of time, environmental changes had forced living organisms to develop particular characteristics in order to survive, and that these characteristics acquired by effort could be inherited. To quote Shaw's simple example: "If you like eating the tender tops of trees enough to make you concentrate all your energies on the stretching of your neck, you will finally get a long neck, like the giraffe." Darwin himself had an uneasy relationship with Lamarckism, in some respects repudiating it and at other times coming close to embracing it. It was not until after the work of Mendel was rediscovered and the science of genetics became established in the 20th century that the idea of inheriting acquired characteristics was proved a fallacy.

Charles Darwin's grandfather, Erasmus Darwin, an 18th-century naturalist and poet, also believed in the inheritance of acquired characteristics. Indeed, some scholars maintain that Lamarck derived his ideas from Erasmus Darwin. In his long poems *The Temple of Nature* and *The Botanic Garden*, and in the footnotes he appended to them, he expounded ideas that his grandson later took into consideration. It is generally recognized that it was from his grandfather that Charles Darwin drew the ideas of sexual selection and the evolutionary significance of the struggle for existence. Erasmus Darwin might have been pre-

Cuvier, Lamarck, and Darwin Senior

Above: Jean-Baptiste Lamarck (1744–1829) from an engraving published in 1805. He was best known for his theory of evolution. Below: *Exhuming the First American Mastodon* by Charles Willson Peale. The first American mastodon to be exhumed, reconstructed, and shown to the public was found in a swamp in Ulster County, New York in 1799, and Peale paid $300 for the remains.

Darwin Voyages to the Pacific

Right: Erasmus Darwin (1731–1802) in an oil painting by Joseph Wright of Derby. He was considered one of the foremost physicians of his day, and is now chiefly important as a transitional figure. In many ways he foreshadows some of the more sophisticated modern scientists like his grandson Charles Darwin and Thomas H. Huxley.

scribing his grandson's life's work when he wrote: "As all the families both of plants and animals appear in a state of perpetual improvement or degeneration, it becomes a subject of importance to detect the causes of these mutations."

Erasmus Darwin was both a gentleman and an amateur in the field of natural science, a frequent combination in 18th- and 19th-century England. William Pengelly was another, an amateur archaeologist who investigated caves in his native Dorset in southwest England and discovered obviously manmade flint tools in the same place as the fossilized bones of prehistoric animals. His find not only pointed to the great antiquity of the earth, but also called into question Cuvier's remark that there is no such thing as fossil man. A similar find was made in northern France, in the Somme valley near Abbeville. The great 19th-century British geologist, Sir Charles Lyell, authenticated these discoveries and declared that they proved man had existed on earth for "immense periods of the past."

Darwin later acknowledged Lyell as the greatest single influence on his own thought, and he dedicated *On the Origin of Species* to him. Lyell's own great work, the *Principles of Geology*, was published in three volumes in the early 1830s when Darwin was an impressionable young man, and these were his constant reading during his five-year voyage around the world on H.M.S. *Beagle*. The *Principles* had a wide readership, and, as well as influencing Darwin directly, it helped prepare the ground for the acceptance of his theories by establishing that changes are brought about by natural forces working over tremendously long periods of time. Despite its title, there is a good deal of biology and zoology in Lyell's *Principles*, as well as a wealth of reflection and speculation. As Loren Eiseley has written, "Lyell possessed in 1830 all the basic information necessary to have arrived at Darwin's hypothesis but did not."

Darwin's study began in 1831 when, at the age of 22, he obtained a post as naturalist on the British survey ship H.M.S. *Beagle*. It was as a result of his observations and reflections during this voyage that Darwin arrived at the central concepts of his *Origin of Species*. Both his diary and autobiography reveal

Above: William Pengelly (1812–94), an amateur British archaeologist. The picture is from an original photograph dated November 20, 1869 and taken in Torquay, Devon.
Left: Charles Darwin, painted at the time of his marriage to his cousin Emma Wedgwood and two years after his voyage on H.M.S. *Beagle*, in around 1838. As an unpaid naturalist on the *Beagle* he had spent five years studying the animal life of South America and Australasia.

how the evolutionary hypothesis gradually gelled in his mind. They also show how he arrived at the two main concepts—the laws of natural selection and of the survival of the fittest—which were the keys to explaining the mechanism controlling the evolutionary process.

Darwin made a series of landings and trips inland as the *Beagle* sailed down the coast of South America, and he was impressed "by the manner in which closely allied animals replace one another in proceeding southwards." Such observations were contrary to the biblical idea of independent creation and suggested the existence of species showing modifications to suit their locality. In Patagonia he discovered the remains of an extinct llama, and this confirmed for him, he later wrote, "the law that existing animals have a close relation in form with extinct

Above: the *Beagle* in Murray Narrow, Beagle Channel in a painting by Conrad Martens. The excitement of such a trip for the young Charles Darwin must have been intense—it was the first time he was to be exposed to such a high degree of intellectual stimulation, which helped inspire his revolutionary and heretical ideas on nature, creation, and evolution.

species." This principle, which he called the "law of succession of types," was important in evolution theory because it showed that new species arose gradually and not by means of special creation. So both his geographical and his palaeontological observations in South America brought home to Darwin the fact that related species branched off from a common ancestry. That was plain to see, but the problem was to explain why they branched off.

A clue was provided in the observations he made in the Galapagos group of islands, which are located on the equator in mid-Pacific Ocean. Here he discovered distinct variations between species inhabiting neighboring islands. On the vast continent of South America variations within species could be attributed to differences in climate and physical conditions, but conditions were the same on all the islands of the Galapagos archipelago. Some other principle must be at work, Darwin deduced, to cause the variations he noted between the clearly differentiated sub-species. His observations in the Galapagos did not solve the problem of the origin of species, but they did rule out certain seemingly possible but in fact false explanations and thus focused his thought on a narrower range of hypotheses.

Not long after his return to England Darwin happened to read the British economist Thomas Malthus' *Essay on the Principle of Population*. In this influential work, published in 1798, Malthus presented mathematical calculations of the growth of populations. He also discussed the consequences of the inevitable growth of populations at a rate exceeding that of the growth of the food supply. Applying the Malthusian analysis to the life of animals, Darwin deduced that they must have had to compete with each other to survive. Clearly those species best fitted to their environment would contribute the most offspring to the next generation, and this generation would have the characteristics that had enabled the previous one to survive, so a species would become progressively better adapted to its environment

The Influence of Alfred Wallace

and thereby predominant within that environment. It could be said that nature acted as a selective force, creating new species from those best fitted to their environment. Thus natural selection and the survival of the fittest, the two key concepts of the *Origin of Species*, were conceived. "Here then I had at last got a theory by which to work," Darwin later wrote, recalling the impetus the Malthusian analysis had given to his thought. But his work at this stage was all done in his head. In 1842 he jotted down a 35-page outline of his ideas, but it was not until 1858 that he took up his pen to write the *Origin*.

He might not have written it even then had not Alfred Russell Wallace forced his hand. Wallace, a British provincial land surveyor turned naturalist and explorer, spent from 1854 to 1862 collecting specimens all over the Malay archipelago. In 1858 he had fallen ill while on the small island of Ternate in the Moluccas, and as he lay in a fever he thought about the problem of species. In a flash the answer had come to him. Later, recalling his sudden insight, he wrote:

"It occurred to me to ask the question, Why do some die and some live? And the answer was clearly, that on the whole the best fitted lived. From the effects of disease the most healthy escaped; from enemies, the strongest, the swiftest, or the most cunning; from famine, the best hunters or those with the best digestion; and so on.

"Then I at once saw, that the ever present variability of all living things would furnish the material from which, by the mere weeding out of those less adapted to the actual conditions, the fittest alone would continue the race.

"There suddenly flashed upon me the *idea* of the survival of the fittest."

Left: Alfred Russel Wallace, Darwin's colleague and co-originator of the theory of natural selection, shown in the Malay archipelago in this modern painting by a Russian artist.

Right: finches of the species *Geospiza strenua*. Darwin wrote: "This singular genus appears to be confined to the islands of the Galapagos archipelago . . . it forms the most striking feature of their ornithology. The characters of the species of *Geospiza* . . . run closely into each other in a most remarkable manner."

Below: the heads and beaks of four different Galapagos finches. This illustration, by R. S. Pritchett from a later version of Darwin's *Journal of Researches* (1889), shows *Geospiza magnirostris* (1); *G. fortis* (2); *G. parvula* (3); and *Certhidea olivacea* (4).

Wallace spent two days developing his idea and writing it out. Then he sent his work to Darwin, with whom he had corresponded in the past. Darwin was astonished to read the very ideas he had been thinking over and harboring for 20 years, clearly expressed by another man. "I never saw a more striking coincidence," he wrote; "if Wallace had my manuscript sketch written out in 1842, he could not have made a better short abstract!" The coincidence forced him to get down to work. He quickly wrote *On the Origin of Species* and published it the next year with the encouragement and approval of Wallace, whose conduct in the circumstances he described as "generous and noble."

The ideas of natural selection and survival of the fittest did not fully explain the great variability of sub-species and individuals within species that Darwin had observed in the Galapagos. To meet this difficulty, Darwin formulated his Law of Divergence. "The same spot will support more life," he wrote, "if occupied by very diverse forms . . . And it follows . . . that the varying offspring of each species will try (only a few will succeed) to seize on as many and as diverse places in the economy of nature as possible." The Law of Divergence, however, was little more than a statement that great diversity existed within species, and did not explain the mechanism that produced it. Darwin kept working at this problem, even pursuing his research in bars, he said, talking to pigeon fanciers.

Again Wallace had an idea. In 1866 he wrote to Darwin pointing out that people had interpreted his doctrine of accidental improvement through natural selection to mean "that favorable variations are rare accidents, or may even for long periods never occur at all." He recommended that Darwin should oppose this interpretation and "constantly maintain . . . that variations of every kind are always occurring in every part of every species, and therefore that favorable variations are always ready when wanted." It was not until the principles of genetics became known that Wallace's insight was proved right, and the mechanism that produced these variations was discovered. It was ironical that the year before Wallace wrote these words to Darwin, Gregor Mendel had announced his findings on genetic laws to the uncomprehending members of the Brunn Society for the Study of Natural Science. The Moravian monk held the secret that the two distinguished English naturalists were trying vainly to discover.

The success of *On the Origin of Species* with the public and the controversy it caused were embarrassing to Darwin, but now that he had gone public there was no turning back, no avoiding the question of what evolution theory implied as to the origin and history of man himself. In later editions of his book, Darwin boldly added a word to his final sentence, so that it read: "*Much* light will be thrown on the origin of man and his history." But when his champion "Bulldog" Huxley published his own book *Man's Place in Nature* in 1863, he claimed with characteristic outspokenness: "Whatever system of organs be studied, the comparison of their modifications in the ape series leads to one and the same result—that the structural differences which separate Man from the Gorilla and the Chimpanzee are not as great as those which separate the Gorilla from the lower apes." So man was undoubtedly one of the great apes. The fact was profoundly shocking to many sensitive and spiritual souls, and the wife of the Bishop of Worcester earned herself a place in any anthology of unintentional wit with the statement: "Let us hope it is not true, but if it is, let us pray it will not become generally known."

Some critics of evolution theory tried to contest it by arguing that, if it were true, fossil remains should exist of species linking man with the apes, and no such remains had been found. Darwin replied that they hadn't been found because they hadn't been sought in the right places. Ernst Haeckel, the late-19th-century German biologist and evolutionist, supported Darwin's view, and he wrote: "The ape-like progenitors of the human race are long since extinct. We may possibly find their fossil bones in the Tertiary rocks of Southern Asia or Africa." Haeckel's prediction turned out to be amazingly accurate, as we shall see, but even at the time he wrote evidence had been found in his own country of the existence of an early man with some apelike characteristics.

The first remains of what became known as Neanderthal man were found by engineers quarrying in the Neander valley near Düsseldorf, Germany in 1856. They gave the bones to the president of the local naturalists' society, Johann Fuhlrott, and he had them authenticated by an anatomist in Bonn, Hermann Schaafhausen, who declared that he was impressed by the "prominent

The Prediction of Ernst Haeckel

Below: Ernst Haeckel, the late-19th-century German biologist and evolutionist, who supported Darwin's guess concerning the whereabouts of then-undiscovered fossils of prehistoric man. He predicted that the link between ape and man would probably be found in southern Asia or Africa. Here he is photographed later in life, holding the insignia of his profession.

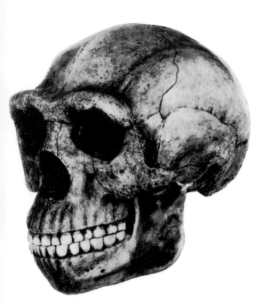

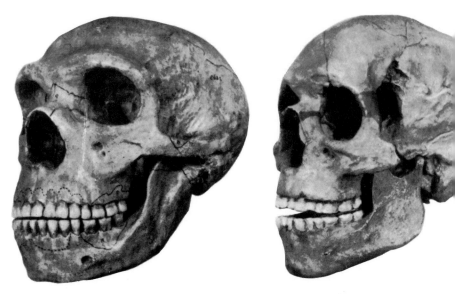

Above: the skulls of (left to right) Peking Man, Neanderthal Man, and *Homo sapiens*. It has been discovered that, contrary to expectations, Neanderthal Man is not the direct ancestor of *Homo sapiens* but rather a separate evolutionary branch. Peking Man however (along with Java Man) is a form of *Pithecanthropus*—"Ape-Man"—which may have existed midway between the ape and modern man.

forehead which in fact is similar to that of a great ape." Thomas Huxley, when he saw the skull, also said that it was "the most ape-like human skull that I have ever seen." Controversy raged over the find, and critics argued that so-called *Homo neanderthalensis* was just a deformed human being such as might be found in the world at any time. This criticism was unanswerable at the time, but hundreds of similar finds in the 20th century have since established that Neanderthal man and his culture did indeed exist, although they were not the direct ancestors of the present *Homo sapiens* but a separate evolutionary branch which eventually became extinct.

The search for the "missing link" between man and ape has been pursued by archaeologists and anthropologists from the mid-19th century to the present day. Anti-evolutionists questioned, criticized, and ridiculed every discovery until the evidence reached overwhelming proportions. The outstanding find of the 19th century was the "Java Ape Man" discovered by a young Dutch anthropologist, Eugène Dubois, in 1891. This was a triumph for Ernst Haeckel, who had not only predicted the area where transitionary man would be found, but had also correctly predicted the characteristics of Java Man and given him the name *Pithecanthropus* (which means simply "Ape-Man"). The relevant characteristics were a brain capacity of about 800 cubic centimeters—that is, midway between the size of the ape's brain and modern man's—and evidence of upright posture, as shown by the shape of the thigh bone. Experts assembled at a conference in Berlin in 1895 examined the fossilized bones and were, with a few exceptions, convinced that *Pithecanthropus erectus* was a transitional ape-man and constituted evidence for the evolutionary hypothesis. Further *Pithecanthropus* finds in the same area in the 1930s, together with stone implements, clearly established that Java Man was a genuine early form of man. Remains with the same characteristics, which were designated Peking Man, were found in China, which suggested a widespread distribution of the type throughout eastern Asia about half a million years ago.

That was longer ago than the evolutionists had imagined man to have been on earth, but finds of even greater antiquity came to

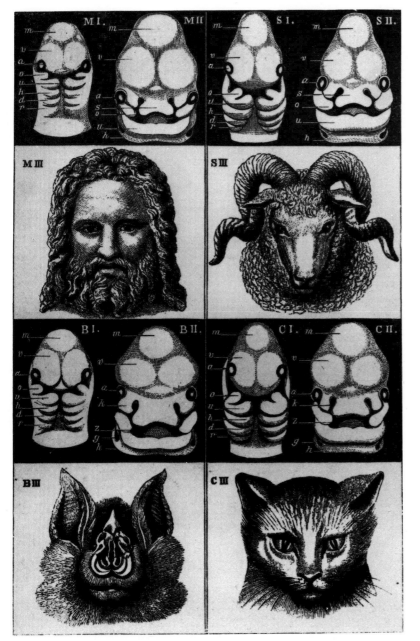

The Search for the "Missing Link"

Left: development of the face in mammals (man, bat, cat, and sheep) at three different stages. This illustration, from *Evolution of Man* (1879), shows Haeckel's idea that the ontological steps in mammals are very similar.

light in Africa, the other geographic area where Haeckel had predicted that fossil man would be found. In 1924 an Australian anthropologist working in South Africa, Raymond Dart, discovered a skull embedded in limestone in a quarry at Taung, just north of Kimberley, which was estimated to be a million years old. It was the skull of a child aged about five years. The characteristics that persuaded Dart it was human were the size of the brain cavity, the formation of the teeth, and the position of the *foramen magnum*—the hole by which the spinal cord enters the skull—which indicated an upright posture. Dart's claims drew the usual skepticism and ridicule, but he was supported by one authoritative voice, that of Robert Broom, a South African who had done world-famous work on fossils of transitional forms between reptiles and mammals. Dart had called the Taung find *Australopithecus africanus* (Southern Ape of Africa), and in the 1930s Broom unearthed skulls and other remains of adult Southern Apes with the same humanoid characteristics as the Taung child. Eventually more than 100 Australopithecines were found in South Africa, and Dart's claims were vindicated. Mean-

Evidence for Darwin's Theory

while he had been engaged in other investigations. He had noticed that many baboon and other animal skulls found in the area of the Australopithecines had fractures indicating death by violence, and extended studies showed that *Australopithecus*, who had lived at least a million years ago, had indeed used weapons to kill animals, and moreover that he had often murdered his own kind.

After World War II investigations of African prehistory continued in the equatorial region of what is now Tanzania in East Africa, where the British anthropologists Louis and Mary Leakey excavated the Olduvai Gorge on the eastern side of the Serengeti Plain. There they found Australopithecines which, according to the new radioactive carbon-14 dating method, were two million years old. In 1961 their son, Jonathan Leakey, found a skull 1.8 million years old with a brain capacity considerably larger than *Australopithecus*, and he became known as *Homo habilis* (the Handy Man) because a variety of tools were found near his remains. This find, along with other similar ones, meant changing the picture of a direct, unbranching line of descent from primitive ape to modern man in which evolutionists had formerly believed. It seemed instead that different species of near-man had existed side by side in the same period, and with the accumulation of more fossil evidence the problem of determining which was the forefather of modern man became more complex. So did the question of when these various hominid species had first appeared in the world. In 1972 a French-American anthropological team found the remains of a female *Australopithecus* which they dated as three million years old, and in the same year the Leakeys'

Right: Richard Leakey, archaeologist and son of Louis and Mary Leakey, holding a fossil skull known as number 1470. Leakey's discoveries over the last two decades have helped to mold evolution theory concerning the descent of man.

younger son Richard found a specimen of *Homo habilis* of the same age with a brain capacity of 800 cubic centimeters, the same size as Eugène Dubois' *Pithecanthropus*.

Richard Leakey's discovery was made in the Lake Turkana region of Tanzania, a site he had had a hunch he should investigate when he happened to fly over it in 1963. His intuition was again rewarded in 1975 when a fossil skull of 900 cubic centimeters' capacity was found. This *Homo erectus* resembled Peking and Java Man, but had lived on the African plains a million years before them.

As a result of these and many other discoveries, evolution theory today holds that the original hominid ape, *Ramapithecus*, appeared more than 10 million years ago. About six million years ago his line branched into three types, two of which were to become extinct while the third was to evolve into *Homo habilis* about 3.5 million years ago. This toolmaker and hunter coexisted with his cousins the tree apes, but he moved out of the forest onto the plains, where the physical conditions of life gave a survival advantage to the upright, socialized, cooperative, weapon-using, communicating species. By the process of natural selection *Homo sapiens* eventually evolved as king of the primates.

Such is the picture of human evolution that contemporary anthropologists have drawn. There are still many gaps and puzzles in the story, so the work of excavation and study of fossil remains continues. One thing that has been amply proved, however, is the truth of the Darwinian hypothesis of the descent of man, which has held up through more than a century of energetic investigation.

Below: East Rudolf in East Africa (now called Koobi Fora), a site in the area where so many fundamental discoveries of fossil man have been found.

Chapter 6
The Walking Ape

Is man, as many prominent modern anthropologists say, still just a naked ape? Do his primate characteristics influence him more subtly than simply in his use of thumbs and similar brain size? We are now told that not only our hunting and mating behavior, but our social, business, and political structures as well are based on primate patterns. Would a return to a purely "natural" environment require many changes in our outlook? Does the ever-present bureaucracy that governs our lives provide a welcome democratic fairness, or does it merely offer dominant roles to sub-dominant humans? To what extent is man controlled by his apelike characteristics?

Why did the brain of man's ancestor, the ape, suddenly balloon to triple its size? This is a mystery that still puzzles anthropologists. For millions of years the brain of the hominid apes had maintained a constant capacity of about 350 cubic centimeters. Then came *Australopithecus* and *Homo habilis* with average brain volumes of 550 and 750 cubic centimeters respectively, and they coexisted on earth up until perhaps 300,000 years ago. The *Homo erectus* group, such as Java and Peking Man, had brains of 850 to 900 cubic centimeter capacity. Suddenly the capacity of the human brain grew to a massive 1400 to 1600 cubic centimeters, and to this day nobody really knows how or why.

There are theories, of course. Orthodox evolution theory would hold that the large brain was naturally selected because, as a maker and user of tools and weapons, man needed improved coordination of hand and brain. Also, as a hunger cooperating with other members of his group, there was obvious survival value in developing a language and memory-storage capacity. That sounds a reasonable explanation, but there is one glaring fact that invalidates it—man's acquisition of a massive brain made no immediate difference to his way of life. In fact he had the extra capacity for hundreds of thousands of years before ever using it. True, *Homo sapiens* only evolved some 10 to 20 millenniums ago. The period of time since then is incredibly short in comparison with the hundreds and thousands of millenniums that went before. But in this time his brain, which had

Opposite: a chimpanzee, *Pan troglodytes*. The similarities between man and his ape ancestors, all classified as primates, are quite evident—beneath the fur this chimp looks and acts very like a human.

Cosmic Accident?

lain dormant for so long, took him from the cave to the space module and from the spear to the nuclear bomb. What triggered this sudden explosion of intellect is another question, but how man began using a faculty he already possessed is less puzzling than the fact that he had evolved it and not known what to do with it in the first place. Nature doesn't usually work like that; characteristics are selected for their immediate survival value.

Whenever orthodox scientific theory fails to explain a phenomenon, intriguing unorthodox theories are always competing for the honor. In 1971 a German writer, Oskar Maerth, brought out a highly original one. "The first man," he wrote, "was that ape which for the first time ate the fresh brain of one of his fellows. The first human beings had become cannibals through hunger for sex. Cannibalism and human development started at the same time: cannibalism is the cause of human development. The first cannibal apes could not know at the start that eating brain not only stimulated them sexually, but increased their intellectual capacities as well. It was not until later that they discovered the effect on their intelligence . . . After eating raw brain there was an immediate effect on the sexuality, but this quickly wore off. These short-lived sexual impulses caused man constantly to renew his warring expeditions against his fellow men, in order to satisfy his greed for sex through eating brain; and this simultaneously led to a lasting growth in intelligence."

Maerth cites as evidence for his theory the fact that most of the hominid ape, Australopithecine, and *Homo habilis* skulls found so far had been cracked open like nuts, as if to gain access to the brain. He also claims that the biblical Genesis story is a symbolic description "of the abnormal development of a hairy animal into man; who, by eating the fruit of knowledge, has become naked, sexually disordered and intelligent." It is an interesting theory, though rather dubious because it appears to hinge on what Dr. James McConnell has called the "Mau Mau hypothesis" that intelligence can be eaten. McConnell, an American researcher specializing in the physical bases of memory, has demonstrated that cannibalism can result in memory transfer in lower organisms. However, he believes that the digestive systems of higher organisms are too efficient for "memory molecules" to remain intact. But of course nobody has really tested the "Mau Mau hypothesis" by including fresh raw human brain in his regular diet under scientific observation.

Another theory explaining man's acquisition of a massive brain as an accident is that put forward by the American writer Robert Ardrey, though with the acknowledgment that there are "mighty objections" to it. What he calls the "Ardrey Theory of Man the Cosmic Accident" holds that conditions favorable to a high rate of genetic mutation were created on earth as a result of its collision with an asteroid or some other large celestial object about 700,000 years ago. Geologists know that reversals of the planet's magnetic field have occurred at times in the distant past, and Ardrey claims that this collision resulted in such a reversal. When these reversals have occurred, a transitionary period has followed of about 5000 years during which the earth has had no magnetic field at all. It is this field that provides protection from cosmic rays, which are known to be mutagenic.

Thus a high mutation rate would have operated during those years, and a big-brained ape could have been the result.

Fascinating though Maerth's and Ardrey's theories are, they are unprovable. However, they do draw attention to two important points: that man has only possessed the large brain making him superior to all other creatures for a comparatively short period of time, and that for the greater part of that short period he did not make use of his singular gifts, but continued to live the life of an ape of the grasslands. The new part of the brain, the *cerebral cortex*, is where those thoughts and activities that we consider uniquely human come from. It has only been in business for a few million years, and although we may think that the instructions of the new part of the brain rule our lives, it is un-

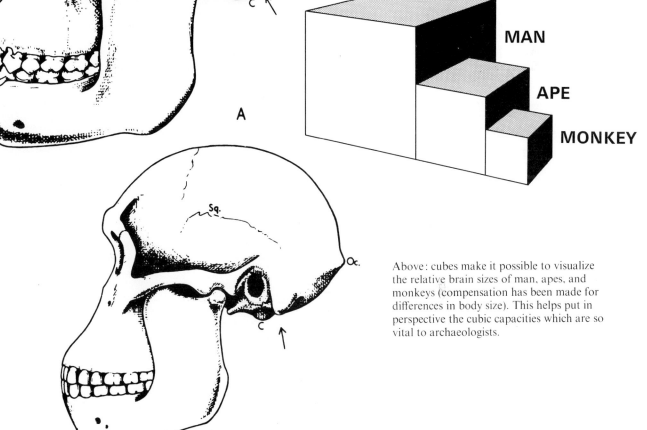

Left: skull of a female gorilla (A) compared with (B) an australopithecine human skull (*A. africanus*). Interesting points are the relative positions of the occipital condyle (c) and occipital protuberance (Oc) at the back of the skull. The jaw is less prominent in B, and the canine teeth do not project as in A.

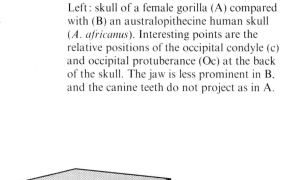

Above: cubes make it possible to visualize the relative brain sizes of man, apes, and monkeys (compensation has been made for differences in body size). This helps put in perspective the cubic capacities which are so vital to archaeologists.

"Monkey Boys" and Naked Apes

realistic to imagine that the instructions and instincts laid down in the old brain over tens of millions of years can be completely overruled by the cortex.

In recent years a number of literary anthropologists, notably the Briton Desmond Morris in *The Naked Ape* and *The Human Zoo*, the American Robert Ardrey in *The Territorial Imperative* and *The Social Contract*, and the Canadian Lionel Tiger and Briton Robin Fox in *The Imperial Animal*, have argued that much of man's behavior and many of his troubles can be better understood if we admit that man is basically simply a species of primate. He is programed to behave like all other primates, and has a relatively small complement of what are known as "species specific" characteristics. These may be the ones he most values and tries to keep in the forefront of his personal and social life, but they are not the characteristics that ultimately govern and define him. Unless man admits this fact he is likely to become very confused. This argument is undeniable up to a point, but the question of just how apelike human behavior is and con-

Right: the "ape boy" of Burundi. One of a handful of survivors of a central African tribal war in 1972–3, he is believed to have been adopted and raised by a band of monkeys. He was found when a group of villagers noticed one of a band of monkeys that seemed slower than the others to climb out of reach of their hunting party. When they caught the creature he turned out to be a tiny, very hairy boy about four years of age, who could not speak and walked on all fours. He still communicated, even several months later, by violent facial grimaces and wild gestures. He has lost much of his previous body hair but still reacts with heavy scowls, and he loves soft food. Anything he is given which requires chewing he immediately spits out, despite having a fine set of teeth. He has been christened John after John the Baptist, the Christian saint who himself lived for a time in the wild.

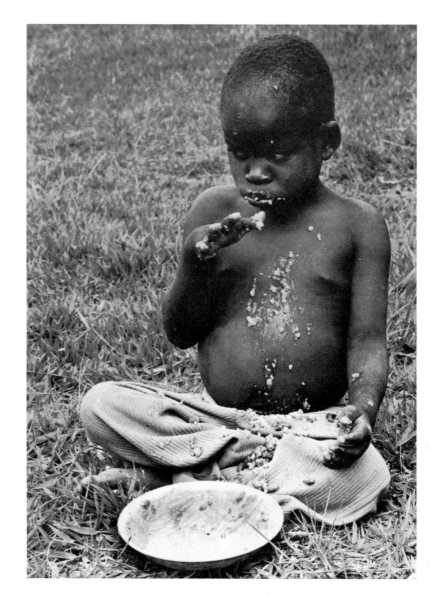

versely to what extent the higher centers of the human brain govern our conduct and institutions is debatable. Objective judgment of the matter is very difficult since the tendency to mock man's faith in his loftier motivations can come from an attitude as self-satisfied as the mocker tries to make those same beliefs appear. In this sort of argument, the conceit of the "realist" must be taken into account as much as the pretensions of the idealist in assessing their respective explanations of the facts. But undoubtedly these modern writers and anthropologists have thrown some useful light on the human animal by comparing it with its primate ancestors and cousins. The fact that Morris' two books alone have sold several million copies around the world must indicate that his insights into human nature are both new and persuasive.

The most conspicuous difference between man and all other species of primate is that he has much less body hair. This is where Morris got his title *The Naked Ape*, and in this book, first published in 1967, one of the first tasks he undertakes is to explain why the process of natural selection favored the evolution of a primate with an exposed skin. Body hair helps mammals maintain a constant body temperature and protects the sensitive skin, and the characteristic would not have been discarded in the evolutionary process without good reason. Several possible explanations have been suggested: that when man's ancestors ceased to be nomadic and became home-based they shed their body hair to help reduce infestation of their permanent sleeping areas by parasites; or that with the discovery of fire our ancestors no longer needed a hairy coat for warmth; or that for a period in its evolution the animal that became man took to the water like those other mammals the dolphins, whales, and porpoises. Morris concedes that there may be some truth in these theories, but he himself favors other explanations. He maintains that the loss of body hair, combined with an upright position which exposed the genitals and the female breasts, worked to increase powers of sexual signaling as well as sensitivity to erotic sensation in sexual encounters. This was important because intense sexual experience tightened the bond between males and females, which was essential to the stability of primate society. Furthermore, man's ancestors lived by the hunt, and body hair would often have caused overheating and consequent exhaustion. Efficiency and stamina in the chase would have been increased by shedding of the hair and increasing the number of sweat glands to help in cooling. With these explanations of human hairlessness, Morris introduces his basic argument: that man is a highly sexed, predatory primate, and "that in becoming erudite, *Homo sapiens* has remained a naked ape nevertheless; in acquiring lofty new motives, he has lost none of the earthy old ones," and "he would be a far less worried and more fulfilled animal if only he would face up to this fact."

Looking back over our evolutionary history, Morris maintains that our ancestors were a species of ape that came out of the forests and took to the plains, where they supplemented their vegetarian diet with animal protein. The females remained the traditional foragers while the males began to hunt in packs. Unable to compete with the other meat-eating beasts on their

Above: a seven-month-old baby chimpanzee brought up by a human parent. Sissi was born in Sweden's famous Kolmården Zoo, but her own mother refused to have anything to do with her. Per Ramland, a zoo official, took her home and raised her like a human baby, but sooner or later she must return to the zoo and live among her own kind. Readjustment will be necessary on both sides.

own terms, the new predators developed the advantageous characteristics they already had. Their upright posture, sharp vision, swiftness of movement, and strong grasping hands were all major assets; but the greatest asset of all was their ability to act in unison, to coordinate the actions of individual members of the pack.

Control and coordination became necessary in the social life as well as the hunting life of our ancestors, particularly after they established permanent homes. Male-female pair-bonding became necessary. This was for two reasons, first to minimize sexual rivalries that might affect the efficiency of the hunting team, and second to give the young a lengthy period of dependence in which to develop physically and undergo learning experiences. A strong parental pair-bond gave the young both maximum security and two experienced teachers. Means had to be evolved for what ethologists, who study animal behavior, have called "sexual imprinting." This is more commonly known as falling in love, and the simplest way to achieve this, Morris suggests, was "to make sex sexier." As part of this process the female became available for sex virtually at all times instead of only seasonally. In the case

Below: young humans often show behavior similar to that of their primate cousins, the monkeys. Not only tree-climbing but chattering, mimicry, and curiosity are characteristic of both species.

Below right: chacma baboon infants playing on a tree branch. They do bear a remarkable resemblance to a certain other species—don't they?

Comparison of Man and Ape

Left: a female crab-eating macaque monkey looking after her babies in the zoological gardens of Basel, Switzerland. Monkeys are well known for the great devotion they show their young, and indeed they make very tender and attentive mothers. They cuddle, groom, feed, and protect their infants, and frequently their attitude towards the males of the group depends upon the treatment their children receive from them.

of the human species she also changed biologically, developing red fleshy lips similar to the labia of her genital region and conspicuous breasts that in turn echoed the alluring roundness of the buttocks. As Morris points out, this breast shape is ill-suited to feeding infants, making it difficult for the child to breathe and feed simultaneously, so it must have developed chiefly as a sexual stimulus. This increased sexiness of the female did more than strengthen the tendency to make long-term pair-bonds, however; it also gave the female power. Female monkeys frequently display themselves sexually to a dominant male not to invite copulation but to arouse him sexually just enough to suppress his aggression. The use of sexual display by females, both to lessen male aggression and as a weapon to gain specific advantages, is of course not unknown in the human species, and when it occurs it is pure primate behavior.

"If one took a group of 20 suburban families and placed them in a primitive sub-tropical environment where the males had to go off hunting for food," Morris writes, "the sexual structure of this new tribe would require little or no modification." This is because "behind the façade of modern city life there is the same old naked ape. Only the names have been changed: for 'hunting' read 'working,' for 'hunting grounds' read 'place of business,' for 'home base' read 'house,' for 'pair-bond' read 'marriage,' for 'mate' read 'wife,' and so on." This analogy between human and primate behavior in the contexts of modern social life, business, and politics is another insight that ethologists like Desmond Morris have made into the human situation. Professor Robin Fox and Dr. Lionel Tiger, in their book *The Imperial Animal* published in 1972, developed Morris' theme with a style and enthusiasm which make their argument very persuasive. "We remain Upper Palaeolithic hunters," they wrote, "fine-honed machines designed for the efficient pursuit of game. Nothing worth noting has happened in our evolutionary history since we

Above: human mothers show many attitudes characteristic of monkeys towards their offspring. Grooming, feeding, carrying, and disciplining their children are activities common to both species. The bond between human mother and baby—one of the strongest emotional ties in human society—seems to be duplicated to an uncanny degree in the social structures of other primates.

left off hunting and took to the fields and the towns—nothing except perhaps a little selection for immunity to epidemics, and probably not even that. 'Man the hunter' is not an episode in our distant past: we are still man the hunter, incarcerated, domesticated, polluted, crowded, and bemused.''

Primate societies are structured as a hierarchy. A typical baboon troop with 40 or 50 members will have a central rank of five or six dominant males around whom the females and the young are clustered. All the other males—including the older ones who have failed to reach high positions in the hierarchy as well as the mature younger ones who are still candidates—are kept at the edges of the group. This makes them vulnerable to attack by predators, but at the same time gives the young opportunities to show their courage and thus strengthen their claims to leadership.

If the baboon troop is attacked or in danger of attack, the dominant males will move out from the center to drive off the predators. At other times their function is to lead the troop and maintain order within it. They will quickly put a stop to any quarrel or fight that breaks out, and they rarely quarrel among themselves. Occasional differences within the hierarchy may be settled by several of its members joining forces to put down the

Below: politicians using primate dominance techniques. Because of his height General Charles de Gaulle of France could easily put Britain's Harold Wilson (left) and George Brown at a psychological disadvantage even before a word had been spoken.

Below right: the commanding presence of a male lowland gorilla. Through his size, aura of menace, and potential strength this primate can establish a dominant position over subordinate males, females, and even perhaps human observers.

minority, but they do this not by violence but by behaving in a ritual threatening manner.

The members of the central hierarchy enjoy priority but not exclusive rights in mating with the females. When a female is in the fertile period in the middle of her monthly cycle she will be monopolized by her dominant male, but when her period of ovulation is over he will allow males from outside the inner circle to mount her. "Everyone copulates; only dominants propagate," as Fox and Tiger put it. To lessen the frustrations of these peripheral males obviously helps to keep the group united, but this system also insures by natural selection that it is the characteristics of the dominants which will be passed on. As in royal and aristocratic human societies, mating has to do with genetics, not with enjoyment. The primary benefit of dominance is a license to breed; the enjoyment of greater freedom, better food, more space, and more confidence than the inferior males are merely secondary rewards.

Dominance is not reached or maintained, however, purely by brute strength. Primate societies are political structures in which qualities of cooperation and intelligence are valuable. Furthermore, the females cannot be ignored. There is a ranking system among them, too, in which some enjoy the privilege of keeping their own young with them long after the others have had to join the peripherals. This makes their offspring a kind of aristocracy enjoying easier access to the hierarchy than other peripherals have. Dominant males must canvass the female vote, which will depend largely upon their past conduct towards the infants of the community. In the baboon troop this means the dominants must maintain an indulgent attitude towards the young, whereas in more complex human societies electioneering politicians sometimes get away with only going through the motions, for instance by kissing babies.

Another striking similarity between primate and human politics lies in what is known as the "attention structure." Dominant males are the focus of attention, and sub-dominants will sometimes neglect everything else just to stare at them. They have what the 19th-century German sociologist Max Weber called "charisma," a quality which also characterizes successful aristocrats and dictators in human societies. This quality is strengthened by elaborate ceremonials and rituals in which these "dominants" play a central part and which focus attention upon them. Modern human politics have become increasingly bureaucratic because of the complexity of the problems they deal with, but this tendency is utterly alien to primate politics. Primates relate to each other "eyeball to eyeball," so it is the dominance of the individual and not his rank that counts. Bureaucracy, however, is a system frequently despised by human beings except for the bureaucrats, who could be considered peripheral males exercising and enjoying a dominance that is not natural to them. Even in the most advanced democracies, contests for status still take place between natural dominants. These may be masked by calling them conflicts over ideology or policy, but the contestants themselves know that the one capable of attracting and holding the most attention has a tremendous advantage over the other, whatever the ideology or policy, and that is pure primate politics.

Dominant Males

Above: power can be expressed through symbols as well as simple brute strength. Here the British Lord Chancellor arrives at Westminster Abbey for a service which precedes the opening of the Law Courts. Attendants carrying the ceremonial mace and purse walk before him. Dominant males can use the "attention structure"—the trappings of office and authority—to focus the eyes of ordinary mortals on themselves. This reinforces their dominant position.

Groups, Packs, Hierarchies

Right: freemasonry is an example of a human institution based on male bonding. This is an early representation of the ceremony raising a member of the brotherhood to the Third Degree. The rite is a reenactment of the murder of Hiram Abiff, King Solomon's mythical master builder. The initiate lies in a mock grave where he will be "stabbed to death."

The system of aristocracy and male dominance may not be the most ideal forms of government. But, write Fox and Tiger, "they have a lot going for them that lies deeper than reason and ideology. Any kind of egalitarian system suffers from a disadvantage. Inequality between men and a division of political function between men and women flow naturally from the nature of the primate that we are." This is not to say that we have to submit to the situation, but if our cerebral cortex urges us to pursue ideals, we must realize that we will have to strive very hard to achieve them. The behavior that was naturally selected over millenniums to insure the efficiency and integrity of our ape ancestors' predatory troops may not be appropriate to human life today, but it is still in our genes. It would be unrealistic for anyone to consider taking up an ideology or ambition without allowing for this fact.

Another primate characteristic of human societies is the existence of institutions based on the male bond. Male-female bonding, as Morris pointed out, is strengthened in humans by prolonged courtship and the refinements of sexual attraction, but in certain circumstances the male-female bond must be put second to the exclusively male bond. Our hunting ape ancestors worked in packs, and it was essential to their success and therefore to the survival of the entire community that the pack should be well coordinated and without friction between its members. This also applied at a later evolutionary stage to groups performing military functions against enemies. So important are these male functions, and yet on the other hand so profound is the male-female bond—a situation which can sometimes cause conflicts between the two sets of demands—that male-bonding must often resort to drama, mystery, impressive ritual, or crude defamation of the female in order to assert its preeminence. For instance, this can be seen in primitive initiation rites, dramatic commitments to blood brotherhood, the mysteries and rituals of freemasonry, the exclusivity of men's clubs, and in the coarseness of a stag party on the eve of a man's marriage.

When metaphors of the chase are used to describe the conduct of business, as for instance when someone speaks of making a "killing," the similarity between primitive hunting and what men now call work stands out clearly. Many primate characteristics can also be seen in the functioning of large companies. There is a hierarchical structure, in which the older, dominant males command the attention of all their subordinates and the peripheral younger males jostle for a higher place, alert for any opportunity to oust one of the established elders. The rewards of dominance are extended living space and enjoyment of creature comforts, including a choice of females. All dominants may not avail themselves of this last reward, but the work of many novelists and filmwriters confirms that as a fantasy it carries strong appeal. Of course the situation sometimes exists indirectly, as in the behavior of a boss who is overlord of a typing pool.

The dominant males in a company—the senior executives—may enjoy the satisfactions of the business "chase" and the "kill," but in modern societies for every male who is fulfilled in this way there are scores for whom the system offers no oppor-

Below: soldiers in training on the assault course at Pirbright, Surrey, in southern England. Society's demand for men to perform military duties requires that male bonds occasionally supercede male-female and other ties. Boys' clubs and exciting stories about war heroes contribute to the preparation of young males for this role.

Above: nursing is a profession created by the attitude of cooperation and caring in modern human society. When one person helps to groom another it can show the mutual concern which lies behind many aspects of primate behavior.

Right: chimpanzees grooming each other. Stroking, brushing, and picking off parasites all reflect the concern and affection which form the basis of a tightly-knit community.

tunity for gratification of their predatory instincts. Wage laborers —the factory "hands" created by the industrial revolution of the 18th and 19th centuries—were not only alienated from the fruits of their labor (as Karl Marx wrote) but also, as Fox and Tiger point out, "from the roots of their biology." They were genetically programed as predatory primates but were deprived of any field of action. Perhaps this is one reason that they formed themselves into aggressive trades unions, and that the younger ones often formed gangs with their own hierarchies, territories, and predatory activities.

Modern human societies are composed, in addition to heavy manufacturing industries, of a variety of "service" industries, and in these, too, analogies exist to primate behavior. Apes spend a great deal of their time grooming each other—that is, picking parasites out of each other's fur—and the mutual give and take, courtesy, and concern involved in this activity are characteristics found in many human encounters. Fox and Tiger suggest that for man, "grooming . . . is the whole range of social activities that have to do with the well-being and being well of the community," including medical care, education, social services, and all forms of social interaction and entertainment, from gossiping over the garden fence to attending the opera. "The basic caretaking program is there, embedded in social signals and social reactions; it took humans to turn it into clinics and charities, into shamans and surgeons, into schemes for public welfare and private recuperation." Primates cooperate to keep each other healthy, and man has taken the further step of cooperating to restore the sick to health and to keep alive the physically weak and handicapped. This difference is significant, for it means that in man the process of natural selection has not exclusively favored the fittest in the sense of the physically healthiest, thus narrowing down the range of possible variations in the gene pool. Instead it has preserved the brainy as well as the brawny, a good thing considering that the challenges facing man in the future will surely demand more from his intelligence than from his physical strength.

Insights into the primate characteristics of human beings, their behavior, and institutions, tend on the whole to be gloomy be-

cause they focus attention upon the unrepentant "beast in man." But this approach does not do full justice to either man or his primate cousins. Primate studies have undergone a revolution in recent years. For decades the views put forward in 1932 by South African-born Sir Solly Zuckerman in *The Social Life of Monkeys and Apes*, a book based on studies of the animals while in captivity and portraying them as sex-crazed degenerates, were widely held by zoologists and general public alike. But studies of the animals in the wild made over the last two decades, notably by the Dutch zoologist Adriaan Kortlandt and the Englishwoman Jane Goodall, who both actually lived among wild chimpanzees for a time, have completely changed the picture. Successful experiments have also been made in teaching sign languages to chimpanzees. Primates generally are skillful, socialized, and intelligent animals. Although modern man may have difficulty reconciling certain elements of his primate programing with the requirements of life in modern human societies, he can take some encouragement from the fact that the "beast" in him is not all bad, and that due to his unique cortical brain endowment he has some choice in the matter of what kind of ape he is going to be.

Re-assessing the Primates

Left: *An Al-Fresco Toilette* by the British painter Sir Luke Fildes (1844–1927). One human grooming another reflects not only the mutual concern of individuals but also the hierarchical system by which both human and ape societies are ordered. "Caring" activities are reassurances that the individual receiving attention is valued; sometimes they are even an acknowledgment that the recipient is superior in rank. This is the case with handmaidens, ladies-in-waiting to queens or noblewomen, valets, and butlers.

Chapter 7
Fight or Flight?

How does a lion trainer use the natural instincts of a killer animal to "tame" it into doing his bidding? Can the behavior of rats enclosed in pens teach us anything about our own behavior in modern overcrowded cities? Rock music performs a function first identified by Aristotle as necessary for the release of emotion—and a modern writer adds that this type of cultural interest may keep human beings from becoming barbarians! Could sports or a greater number of Nobel prizes help to defuse aggressive tendencies in man? Threatening behavior, hatred of strangers, and defended territory— all play a part in our mysterious psychological dilemmas of whether to fight or run.

Most of us will probably never need to train a circus lion, but the method is interesting nonetheless. It is mainly a matter of space and distance. Every animal has what is known as its "flight distance"—that is, the distance it keeps between itself and a possible enemy. If an enemy comes within that distance, the animal will first make threatening noises or movements and then either run away or launch an attack. The lion trainer therefore enters this critical space by just a few inches, causing the lion to back away the same distance, ominously snarling. When it is backed up against the bars of the cage and can go no further, the lion will snarl, roar more threateningly, and prepare to attack if the trainer comes any nearer. If at the last moment the trainer places an obstacle such as a box or a pedestal between himself and the lion, the lion will leap upon it in order to attack him, but when the trainer retreats out of the critical flight distance the animal will be instantly appeased and take no more interest in him. Although the audience will see it as an intrepid trainer directing the lion to climb onto the pedestal, the animal will consider that it has driven off a potential enemy. Honor will be satisfied on both sides, and the trainer will be able safely to turn his back on the lion and take a bow.

This information is illuminating for three reasons. First, because it brings to our attention the relationship between space and aggression; second, because it shows that the great killers of the natural world do not release their aggression unneces-

Opposite: a color print of 1894, *Un Barbier dans la cage aux Lions*. The technique of lion-taming depends on the concept of "flight distance"—an animal will feel forced to protect a certain amount of private space, but beyond that critical area it may feel no necessity to notice an intruder. Lion trainers have known this fact, if only intuitively, for many years.

Man the Killer

Right: man's aggressive—even murderous—instinct has been evident for centuries. Here the Egyptian King Den, fifth king of the 1st Dynasty (around 3000 B.C.), attacks an Asiatic enemy. The scene has been preserved on one of the king's ivory labels.

sarily; and third, because it illustrates how easily the natural aggressive impulse can be placated. Any insight into the nature of aggression and the mechanisms that control it is surely valuable for man today, poised as he is on the brink of self-destruction. The questions of whether man is inherently destructive, violent, and evil, and if so, why he is, have occupied philosophers, psychologists, and theologians through the ages. In the light of modern knowledge made up of Darwinian theory and studies of animal behavior, we are today better equipped to answer than people have been in the past. The first question—whether man is inherently destructive—can in fact be answered quite simply, with the statistic that in the last 150 years he has killed about 60 million of his own kind. The mystery, on which the survival of his species may depend, is why man has behaved towards his fellow men with this consistent savagery, unparalleled by any other species.

Our knowledge of evolution theory might suggest an answer. Although man is descended from a species of predator, he is not so obviously equipped for that function as are the great carnivorous predators with their formidable teeth, claws, and great strength. It was man's invention and use of weapons, along with his development of an intelligence enabling him to cooperate, that made him an effective predator. Acquired characteristics are not genetically inherited, but when a certain level of intelligence is reached these traits can be *culturally* inherited, or passed down through custom and teaching. Man's use of weapons is culturally inherited, and put in his hands a power that was not controlled by any instructions from his genetic inheritance. The great carnivorous predators are restrained by powerful blocking mechanisms, particularly when it comes to fighting their own kind. Their threatening behavior, like the circus lion's, can easily be pacified. But man's genes do not carry such strong inhibitory

instructions which would prevent the killing of his own species, because without weapons he is not equipped to be a formidable killer anyway. The relatively new part of his brain, the cerebral cortex, inhibits him to an extent, but the reason and sympathy it exercises over man's murderous tendencies is limited. This fact is attested by that figure of 60 million dead in 150 years, and by consideration of the atrocious ways in which many of those died.

However, there are other considerations to take into account aside from evolution theory. For one thing, existence in an environment filled with hazards that call on aggressive instincts makes them essential to the health and survival of a species. Man's undoubted success as a species is largely due to those instincts. There is also the consideration of space and overcrowding. The last 150 years have seen an explosive growth in the earth's population, so that from an evolutionary and biological point of view the 60 million dead represent some relief of pressure on available space and food supply. It may be relevant to ask whether man's murderous tendency during this period has served the purpose of keeping population down. There are many examples in nature of species keeping their populations constant or reducing them in the face of overcrowding by normally uncharacteristic behavior. There is no doubt that man's murderous tendency is built in. However, there are clearly some circumstances that enhance and other circumstances that suppress the characteristic. Debate continues on whether this tendency is a mechanism of the evolutionary selection process or is due to "original sin." It may even be because, as the Hungarian-born philosopher and journalist Arthur Koestler has suggested, "there is some construction fault in the circuitry that we carry inside our skulls." We may be better able to judge these possibilities if we take into account observational and experimental evidence of the aggressive instincts in animals and men.

Above: war has claimed more victims, sometimes pitting brother against brother, as we have moved into modern times. This scene of the American Civil War shows the 1862 Battle of Fredericksburg (Virginia). General Ambrose Burnside, the Union commander, lost more than 12,000 men in a series of desperate attacks on the Confederates under General Robert E. Lee.

Below: mass warfare develops from an internal dispute between an emperor and one of his subject nations. A British MP guards a wounded German prisoner captured at the Battle of Ancre in November 1916.

The "flight distance" of an animal or person varies between species and to some extent between individuals. Experiments with human prisoners have shown that those with histories of violence begin to show anxiety and fight or flight symptoms at a distance four times as great as that at which normal people react. We all need our own personal living space, a larger space safe from intrusion for our family group, and a still larger one for the nation we belong to. If any person or alien group threatens to invade one of these spaces, our hackles rise and we are ready to fight.

Territory is the first thing animals and men fight over. The "territorial imperative"—the need to have and hold a territory—is intimately bound up with male confidence, courage, pride, and sexuality. According to the Austrian zoologist Konrad Lorenz in his book *On Aggression*, published in 1966, "if we know the territorial centers of two conflicting animals . . . all other things being equal, we can predict, from the place of encounter, which one will win: the one that is nearer home." If the victor pursues a beaten animal towards its home territory, they will eventually reach a point at which the fugitive recovers its courage and strength to the extent that it will turn to drive the pursuer off. It was a birdwatcher who first observed, in a book published in 1920, that female birds were unresponsive to males without territory.

In 1952 the British zoologist Desmond Morris carried out an experiment on a fish—the 10-spined stickleback—which produced some bizarre results. In the species' normal reproductive procedure the male builds a nest of water weeds, entices the female to enter it and lay her eggs there, and then drives the female away. He then fertilizes the eggs and eventually raises the young. In Morris' experiment five males were put in a tank only big enough for two established nests and territories. After a series of fights two dominants emerged, and the other three retired to the corners of the tank. When a female ripe with eggs appeared, a nested male stickleback performed his courtship dance. This entices the female towards his nest; if she hesitates to enter, the male urges her on with a little bite, which may, however, have the opposite effect of driving her away. If she enters, she leaves her tail protruding, and the male excitedly rubs his nose against it, which causes her to lay her eggs. The interesting observation Morris made was that often, when a female had fled before entering the nest, her place would quickly be taken by one of the three sub-dominant males. He would allow himself to be enticed to the nest and would then behave exactly like a female, leaving his tail protruding while the dominant male stimulated it, then swimming through the nest and out the other side. Of course, these males left no eggs for the other to fertilize.

The implication of this experiment was that overcrowding a territory caused the displaced males within it to become psychologically and sexually abnormal. Seventeen years later Morris developed this theme with reference to human societies in his book *The Human Zoo*. "Under normal conditions, in their natural habitats, wild animals do not mutilate themselves, masturbate, attack their offspring, develop stomach ulcers, become fetishists, suffer from obesity, form homosexual pair-bonds

Above: the nesting habits of the 10-spined stickleback fish, studied in detail by the British zoologist Desmond Morris, seem to imply that overcrowding causes displaced males to become psychologically and sexually deviant. Whether this conclusion can be applied to humans is a subject of some debate.

or commit murder. Among human city-dwellers, needless to say, all of these things occur." But this does not necessarily mean there is a basic difference between human beings and wild animals. "Other animals do behave in these ways . . . when they are confined in the unnatural conditions of captivity. The zoo animal in a cage exhibits all these abnormalities that we know so well from our human companions. Clearly, then, the city is not a concrete jungle, it is a human zoo."

In the overcrowded city the male dominance confrontation—the challenge to a rival to fight or flee—often takes the form of a struggle for status rather than territory. The losers in the status struggle, or those who are forced to give up status because of their age, often become prone to what are known as stress ail-

The Causes of Aggression

Below: the 10-spined stickleback in congenial surroundings. It was only when laboratory conditions were made unbearable that members of the species developed abnormal behavior sometimes thought to be solely the province of human society.

ments, such as cancer, coronaries, and ulcers. Status is a less secure possession than territory, and even those who hold it often do so under stress. It is an interesting and significant fact, though, and one which contradicts generally held beliefs, that men with high status and responsibility are less prone to stress ailments than those who are "underdogs." A medical study of 270,000 male employees of the Bell Telephone Company in the United States showed a diminishing likelihood of coronaries in the higher ranks: at one extreme, ordinary workmen suffered from them at the rate of 4.33 per thousand per year, and at the other, top executives succumbed at the rate of 1.85 per thousand. As the average age of executives is considerably older than that of workmen, the difference between the two rates is highly significant and indicates that the health hazard associated with dominance is far less than that relating to sub-dominance. In other words, in the struggle for status in human societies, as in the struggle for territory in animal societies, the loser is an all-around loser. In the animal world deterritorialized males have been known to lie down and die of defeat, and of course it is also common for high-status men to die soon after retiring.

Seeking a physiological explanation of death by stress, the American Robert Ardrey writes: "We believe that subordinated animals experience enlargement of the adrenal gland, and under the pressure of sufficient stress may, through adrenal exhaustion, sink into apathy or death." It is true that the physiological changes a body undergoes in the fight-or-flight confrontation are quite dramatic. Adrenalin pours into the blood, which is drawn from the skin and internal organs to be pumped into the muscles and the brain; the blood produces extra red corpuscles and draws energizing sugars from carbohydrates stored in the body; blood thickening processes speed up. Many internal body functions, such as movements of the stomach and intestines, are hindered, respiration becomes deeper to increase oxygen intake, and sweating—the temperature control mechanism—begins. All these changes prepare an animal to fight, although some of the energy they accumulate may be discharged in preliminary threatening behavior. But if no fight occurs, the body takes some time to return to normal functioning and get rid of the residue of chemical energizers. During a confrontation animals often engage in what are known as "displacement activities," such as scratching or preening themselves or making incomplete feeding movements. As with human beings who may crack their knuckles or take a drag on a cigarette, such activities use up a certain amount of the accumulated extra energy. "Taking it out" on a subordinate after backing down from a confrontation—a behavior pattern common to both animals and humans—may get rid of some more, but constant suppression of aroused but unexpressed aggression ultimately leads to physiological damage. This is shown most commonly in man in the stress ailments, particularly coronaries and ulcers.

Desmond Morris has suggested that the average circle of acquaintances of most city-dwellers approximates the number of individuals who would make up a small tribal group. Thus "in our social encounters we are obeying the basic biological rules of our ancestors." In such a group, numbering between 50 and

Above and below: a confrontation situation between two olive baboons requires that one or the other take some course of action. The fact that one turns to flee and the other to chase may relate to the distance each animal is from his home or territory, the nearer one feeling a greater compulsion to defend it at all costs.

100 individuals, the dominance hierarchy is quickly determined and stabilized. The members will show hostility towards strangers and possibly adopt modes of dress or manners of speech which exclude them. Robert Ardrey has shown in *The Social Contract* that *xenophobia*, the feeling of mingled distrust, fear, and hatred of strangers, is a factor in the life of all organized societies, particularly primate societies. The conditions of modern human life have created a situation in which, as he says, "we must invent strangers." So we get the city-life phenomenon of in-groups and out-groups, which among sophisticates might take the form of shared cultural interests and elite gathering places, while at the other end of the social spectrum it might show itself in antagonistic male gangs commanding well-defined territories. In his native Chicago, Ardrey writes, "the rotting, packed Negro ghetto was in truth a series of separate villages," and gang warfare between them was common.

But the city, "human zoo" though it may be, is of man's making. To seek a remedy for man's problems in a policy of "going back to nature" would be naïve. There are those who thrive on the conditions of city life, who are stimulated rather than sapped by its tensions and challenges. The success of the species *Homo sapiens* is partly due to its flexibility and adaptability to changed conditions, and it is arguable that the challenge of the city might call forth from man a response that will take him one further step up the evolutionary ladder. As Ardrey puts it, "The city forces upon all of us whatever reason man can mobilize." And Konrad Lorenz makes the incisive reflection that "the long sought missing link between animals and the really humane being is ourselves!"

But is there, as Arthur Koestler says, "some construction fault in the circuitry" of the human brain which gives us a perverse urge to crowd each other and deliberately create conditions that must lead to the eruption of violence and destruction? The case of John Calhoun's rats may be relevant to that question.

Calhoun, an American scientist working at the National Institute of Mental Health in Washington, D.C. in the 1950s, devised a series of four interconnecting rat pens which were assembled in such a manner that the two end pens had only one entrance each and therefore could be defended by one rat. The middle pens, however, had two entrances each and were indefensible and open to all. He introduced into this system a rat population of a size which it could normally be expected to support. Predictably, an initial struggle took place between the males, leading to the emergence of two dominants who established territorial rights over the two end pens, where they attracted harems of females. Under these conditions the females could build nests, bear their young and rear them in security. They continued to do so even when Calhoun increased the population of the complex. The two dominants had no trouble retaining control of their territories, but the overcrowded middle pens became, as Calhoun said, a "behavioral sink." A class of dominant males existed there, but none could establish a territory, and they continually fought among themselves. They also attacked males of a middle ranking, who rarely ventured to fight back but nevertheless were rivals for the services of the females. The males of a

Confrontation or Gang Warfare

Above: gang warfare in urban alleyways may also result from the need to defend a territory. An individual's need to prove his physical and mental dominance may also be a strong factor.

Observing the "Criminal" Rats

third class were completely subordinate; ignoring and ignored by the rest, they acted like sleepwalkers and had apparently lost any sexual drive. But the surprise result of the experiment was the emergence, with further overcrowding, of a fourth class, a type of "criminal rat" which sometimes formed gangs and behaved as if sexually obsessed, engaging in homosexuality and even rape. These degenerates would not wait outside a female's burrow in observance of normal rat sexual etiquette, but often forced their way in, savagely violated the female, and killed and ate her young. The analogy to some of the worst aspects of human city life was very clear, but Calhoun made one further observation that was even more chilling in its implications. The secure and protected females of the end pens seemed to find the hurly-burly and excitement of the middle pens, despite its hazards and horrors, irresistibly appealing. They often joined in the melee with every sign of relish, returning after their spree to their own nests in the territories protected by their overlords.

To apply observations of rat society and psychology to human beings is of course a dubious way of arriving at the truth. However, the similarities in Calhoun's rats' behavior to human ten-

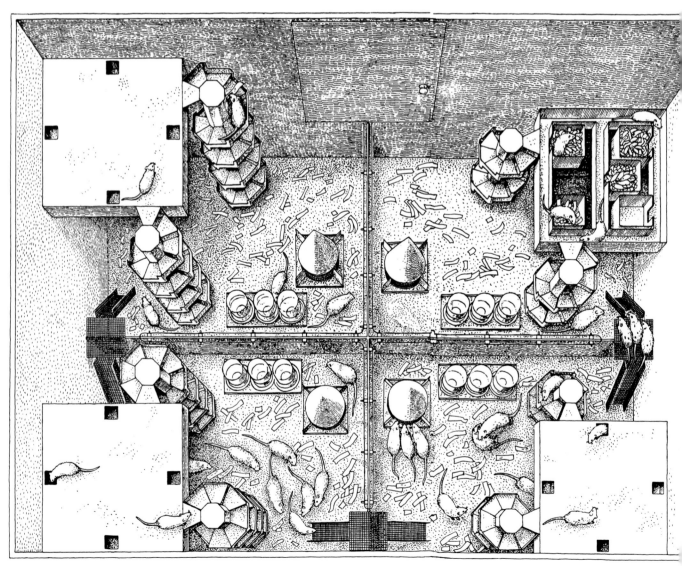

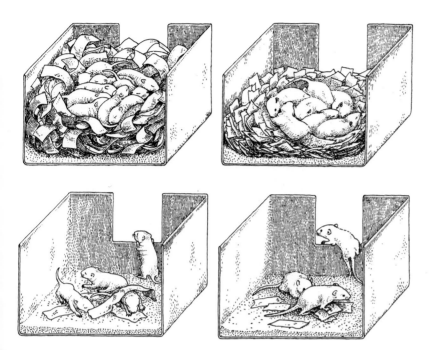

Left: further observations by John Calhoun during his experiments. Normal maternal behavior among rats includes building a proper nest for the young (above left and right). Abnormal maternal behavior, shown by females exposed to the pressures of population density, includes a failure to build adequate nests. In these cases the mortality rate among infants of disturbed mothers was as high as 96 percent.

dencies is startling, and, having seen the behavior objectively and in perspective, it is natural to ask whether humans likewise crowd together in the worst areas of our great cities because the life there has an irresistible attraction for them. That may well be so; man may not be trapped in the "behavioral sinks" of his cities by unalterable social and historical forces, but by his own free choice. It could be argued further that in human societies there are analogies to Calhoun's end pens, where order prevails and the genetic endowment of those dwelling there is perpetuated, in which case the "behavioral sinks" would be functioning as a kind of evolutionary selection device.

In another suggestive experiment with rodents a highly aggressive animal was put in the home of a particularly pacific one, but hardly had the inevitable fight begun than the aggressor was pulled out of the area by a thread attached to its foot. The pacifist thought he had won, and the next time the aggressor turned up he was ready to put up a real fight. After he had "defeated" the other several times, his confidence and fighting spirit were so aroused that he became agressive himself and went about willfully attacking the weaker members of the community, particularly females and juveniles.

The "underdog" deluded into believing he is "topdog" and behaving accordingly until he is put down again is a stock device of theatrical and cartoon comedy. But if aggressive violence lies as close to the surface in humans as it does in rodents, the situation is not so funny. Nor would we welcome the thought that the behavior of the fresh-water cichlid fish, as reported by Lorenz, is analogous to that of human beings.

Of a number of cichlids reared together in an aquarium tank, one adult dominant pair, distinguished from the rest by brighter colors, eventually tries to drive all the others away and take over the entire territory. The defeated swim around the corners and across the surface of the tank. If at this stage the aquarium

Opposite: John Calhoun's experiment concerning the effect of population density on the behavior and social organization of rats. Groups of 80 animals were confined in a 10-by-14-foot room divided into four pens by an electrical fence. Each of the pens (numbered 1, 2, 3, and 4 clockwise from door at top) was a complete dwelling unit. The conical objects are food hoppers, and the trays with three bottles are drinking troughs. Elevated burrows, reached by winding staircases, each had five nest boxes (seen in pen 1 where the top has been removed). Ramps connected all pens but 1 and 4. The rats therefore tended to concentrate in pens 2 and 3. The development of a "behavioral sink" is reflected in pen 2, where three rats are eating simultaneously and one is attacking another. In pen 3 a nursing female is pursued by a pack of males. The dominant males of pens 1 and 4 are sleeping at the base of the ramps so they will wake when intruding males approach their territories. The three rats peering down from the ramp between pens 1 and 2 are probers, a deviant behavioral type produced by the pressures of high population density.

Man's Unique Aggressiveness

keeper removes them in order to leave the happy couple to breed in security, he will find one day that the male has vented his violence upon his mate and killed her, as well as destroying any young or eggs. What that master portrayer of human depravity, the American 19th-century writer Edgar Allan Poe, called "the imp of the perverse" is clearly not an exclusively human characteristic.

The cichlid family tragedy does not occur if a scapegoat is left in the tank for the belligerent male to harass. Nor does it occur —and this is particularly significant—if a glass screen divides the tank into two territories with a pair of cichlids in each half. In this case the fish seem to discharge their aggression by threatening each other through the glass. "Going through the motions" apparently satisfies the needs both of honor and physiology.

Most animals can limit or curtail their acts of aggression against members of their own species through ritual or displacement activities. The purpose of this aggression is to establish and maintain social order, so that once this is accomplished there is no need for any actual fighting. Man is unique among species in the extent and destructiveness of his aggression against his own kind, possibly because his biological evolution has lagged behind his cultural evolution, and he is not equipped with biological mechanisms powerful enough to inhibit his behavior. He has outstripped the destructiveness of other species by inventing weapons. These were initially a means of extending and augmenting the powers of his arm and hand, but today they have become extensions of his mind, in the sense that only a mental act, a decision to give an order or to push a button, is required to unleash weapons of unprecedented destructiveness. In these circumstances the aggressor does not even have the sight of the appalling results of his actions to inhibit him. Even without such power man has proved himself the most efficient and merciless killer in the animal world, which makes the future of *Homo sapiens* look rather gloomy to some observers. So innate human aggression must be taken very seriously, and it should be considered whether our understanding of evolution and increasing knowledge of how other species cope with these problems can teach us anything.

The first thing evolution theory tells us is that it is neither possible nor desirable to eliminate aggression completely. It is as essential to the determination of social order as it is to the acquisition of a sense of individual identity. But the problem seems to be that the human species has an aggressive urge which is disproportionately strong compared to its evolutionary and psychological requirements in the modern world. The general problem is magnified in the specific situation of the Utes, a tribe of prairie Indians in North America. Natural selection had for generations favored the breeding of warrior males, and today, according to the American psychiatrist Sydney Margolin, the Utes suffer more frequently from neuroses and are more often involved in motor accidents per person than any other human group. In a rapidly changing world biological characteristics adapted to previous needs can quickly become obsolete and inappropriate. But since they cannot be shed like a snake's skin, they must be somehow rechanneled.

Left: the steeplechase event at the Olympic Games held in Mexico City in 1968. Can international sports competitions between individuals or teams provide enough of a release of aggression to prevent a third world war?

Below: perhaps one of the most direct forms of aggressive behavior sanctioned by society. Brand of the United States defeats Sasak of Japan in the middleweight wrestling event at the 1968 Olympic Games in Mexico.

Contests,Not Confrontation!

Right: Anthony Storrs, the British author of *Human Aggression*, suggests that the "space race" between the United States and the USSR was a magnificently harmless outlet for the inevitable rivalry between the superpowers. Here the American Apollo 15 spacecraft, with three men aboard, blasts off from Cape Kennedy in Florida on July 26, 1971. The fact that a huge number of people around the world watched Apollo 11 —the first manned landing on the moon, by US astronauts in 1969—supports Storr's point. Could other fields of human endeavor be rewarded by some sort of Super Nobel Prize system?

"Human beings today . . . succumb to barbarism because they have no more time for cultural interests," writes Lorenz. Aristotle long ago recognized that one of the functions of the arts is what he called *catharsis*, or the outlet of emotion. Freud contributed the concept of *sublimation*, or gratification of basic instinctual drives through diverting them into socially acceptable behavior. It will also be obvious to anyone who has heard modern rock music that much of it offers its followers a catharsis or sublimation of aggressive drives, and although this may not be the kind of cultural interest Lorenz had in mind, it is obviously needed and effective. Public support of teams at sporting events is another effective means of rechanneling natural tribal aggression, and in his book *Human Aggression*, published in 1968, British psychotherapist Anthony Storr recommends encouraging the increase of international competitive sport as a way to reduce the dangers of war. He also points out that the "space race" was an excellent means of expressing Great Power rivalry, and he proposes that more trophies should be awarded for competition over a much wider range of human endeavor than presently covered by, for instance, the Nobel Prize system.

But the problems of population density and the anonymity of the city populations among which many people live as well as the company hierarchies under which they work would not be affected by such developments. That visionary British economist, the late E. F. Schumacher, maintained that "small is beautiful" in the organization of industrial and community life. What we know of our evolutionary past tells us that small is not only beautiful but also necessary if human beings are to be confident of personal identity and status. Without the harmless means of discharging their aggressive drives, those drives could become fixed upon racial, ideological, or national objectives. This could result in an irrational destructiveness the like of which the world has never seen. Man must accept and learn to live with the biological imperatives of his species if he is to be, as Desmond Morris puts it, "a far less worried and more fulfilled animal." We might add to Morris' observation the further one that unless man does become less worried and more fulfilled, he is in danger of wiping out a species that, so far as we know, has been evolution's most interesting and promising achievement to date.

Below: a tournament scene painted by the late-15th-century Italian artist Domenico Morone. Such contests of skill and strength between individuals provided a release of aggression in medieval times aside from keeping warriors in training between times of fighting. It is interesting to speculate, however, what the poorer peasants did to release their feelings of aggression. (Reproduced by courtesy of the National Gallery, London.)

Chapter 8
Five Senses-or More?

Our senses link us with the outside world, with each other, and even with our own bodies. Do each person's senses receive the same signals? Why can some people hear or see so much better than others? Are man's senses deadened compared to those of other animals? The workings of the eye, how we differentiate between colors, ultrasonics—scientific investigators are still making discoveries in these fields, through experiments on humans as well as bats, frogs, and alligators. What do lovesick dolphins say to each other? But perhaps the most exciting and mysterious developments are those theories and experiments which explore the hotly-debated subject of a sixth sense!

Twenty small objects were handled by 20 different people then put in a box. A blindfolded girl was brought into the room. She took from the box one object after another, and after smelling each object and the hands of the assembled people she gave each person the object he or she had originally handled. In other tests, the girl demonstrated her ability to locate by her acute sense of smell a grain of musk hidden anywhere in the room, and, when she was brought blindfolded to where several of her acquaintances were gathered, she was able to identify each one at a distance of a yard just by smelling the air.

In an experiment to test her sense of hearing, the same girl was able to hear a sound like the hiss of a snake at a distance of 230 yards, whereas other people could not hear it from farther away than 30 yards. But the girl was not a freak. She was a perfectly normal person acting under hypnosis. When she was not hypnotized, her senses were no more acute than anyone else's.

Her sister showed similar abilities. Under hypnosis she was able to taste quinine dissolved in water in a solution of one part to 600,000, but the solution had to be four times stronger for her to detect the quinine when she was not hypnotized. Again, under hypnosis she could visually identify acquaintances at far greater distances than other people could see, and her sense of touch became so acute that she could unfailingly distinguish between the two poles of a magnet when either pole was touched against her skin.

Opposite: *The Blind Girl*, painted in 1856 by the British artist Sir John Everett Millais. It is often true that with the loss or impairment of one of the five senses, particularly sight, other senses become sharper. Here the girl seems more than usually sensitive to music and perhaps to the feel and smell of the air. (Published by permission of Birmingham Museums and Art Gallery.)

Dr. Nicholas Humphrey of Cambridge
University in England has studied the
artistic sense of rhesus monkeys. He
believes that an aesthetic appreciation must
be widespread in nature and have a survival
value rather than being either exclusive to
man or a "non-essential" quality. Rhesus
monkeys were chosen because their eyes are
structurally similar to humans' and have
exactly the same range of color vision. To
find out which pictures the monkeys
preferred, he put each one into a chamber
with a screen at one end. Upon pressing a
button, one of two pictures was projected.
If the monkey liked what it saw, it could
hold the button down; if not, it could let it
come up and then press for the other
picture.

Right: Mickey Mouse movies were popular.
The monkeys would watch each film avidly
for as long as it ran.

Below: pictures of bananas aroused little
interest, even if the monkey was hungry.

The subjects in these experiments, performed early in this
century, were two South African sisters of Dutch descent, aged
21 and 18 respectively, and the experimenter was the South
African writer and naturalist Eugene Marais. After the Boer
War at the turn of the century Marais had lived for three years in
a hut in the Waterberg mountains, in the Transvaal, South
Africa, in order to study at close quarters the behavior of a troop
of wild baboons. Marais' observations of these animals gave rise
to some new thoughts on evolution, and the experiment with
the Boer girls was designed to test one of his theories.

The senses of sight, hearing, smell, taste, and touch are,
Marais suggested, less acute in man than in other animals because, with the development of his intelligence, man became less
dependent on his senses for survival. Its senses give an organism
information about its environment, information upon which it
immediately acts in order to avoid or protect itself from predators, or to obtain food. But man's ability to learn from experience,
and to plan and prepare for the future, served him better in his
struggle with his environment than the most acute senses ever
could. Therefore natural selection tended towards the further
development of these "intelligence" functions rather than of
the senses. Marais suggested, however, in his book *The Soul of
the Ape*, published in 1969, that "this degeneration in man is not

organic, or even functional in the generally accepted sense of the term. The organs are still capable of a very high degree of sensitiveness, and under hypnosis they may actually become functional." He demonstrated his point in his experiments with the two Boer sisters.

For purposes of comparison, Marais simultaneously tested the acuteness of the senses of chacma baboons. As the chacma is not as highly evolved as man, but as a primate is more highly evolved than other animals, Marais anticipated that its sense-acuteness would come "midway between that of normal man on the one hand and of the higher mammalia on the other." And he was right, at least in the fact that the chacma proved more sensitive than the normal human being in all five sensory functions. Under hypnosis, however, the two girls' senses of hearing, smell, and touch proved even more acute than those of the chacma. Marais' theory that man's high intelligence had been naturally selected at the expense of his sensory functions, and that his sensory acuity could be enhanced by "deactivating" the inhibiting intelligence through hypnosis, appeared well demonstrated.

Perhaps the most interesting part of Marais' theory was its claim that man's senses were potentially more acute than generally believed. Marais proposed an evolutionary reason for the inhibition of the senses, but other scientists since his day who have considered the same phenomenon have instead suggested a biological reason. Sensory systems, they maintain, were evolved to serve the survival needs of an organism within its environment, and they function mainly to reduce and filter information, passing to the brain only information relevant to those basic needs. As the 20th-century British writer Aldous Huxley put it

Sensory Functions

Left: the monkeys showed strong and consistent reactions to colors. Although little could be discovered about their favorite colors, Dr. Humphrey found that all the subjects tested agreed on one point— they all hated red. Humans also react strongly to this color.

in his book *The Doors of Perception*: "To make biological survival possible, Mind at Large had to be funneled through the reducing valve of the brain and nervous system. What comes out at the other end is a measly trickle of the kind of consciousness which will help us to stay alive on the surface of this particular planet."

The most famous scientific experiment illustrating this point was conducted in 1959 by a research team at the Massachusetts Institute of Technology in the United States. It was reported in a paper entitled "What the Frog's Eye tells the Frog's Brain." The researchers implanted microelectrodes in a frog's optic nerve to measure impulses sent along the nerve to the brain, and they put the frog in a position where many different objects, colors, and movements could be presented to it. They discovered that only four types of visual information were sent from the eye to the brain, all of which had to do with the detection of possible predators or food. The eye seemed designed to discard all other information.

Right: the sense of sight (like other senses) in any animal is adapted to its survival needs and its particular habitat. The frog's eye sends its brain only information which will be useful and filters out other stimuli. The frog does not seem to see, or at any rate is not concerned with, the details of stationary parts of its environment (it would starve to death if surrounded by unmoving food). When escaping enemies it seems merely to leap to a darker spot.

Right: a photographic record of the response to a stimuli of one type of fiber in a frog's optic nerve. This group of fibers responds to differences in shade—contrast—between the object viewed and its background. A single fiber reacts to the passing of a black disk across its field of vision (a). The same fiber gives a continuous response (b) when the disk is stopped within the field.

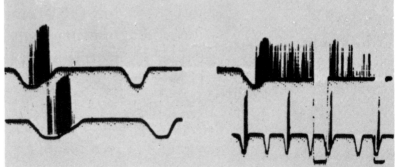

Similar experiments have been conducted with cats and monkeys which produced similar results. Of the five senses, vision is the one most suitable for such experimentation because the visual system is the easiest to observe and stimuli can be presented to it in a controlled manner. But scientists believe that what has been established for vision is true of all sensory systems: they function selectively in order not to pass on to the brain useless or irrelevant information. If they did not do this, the world would be for all of us what the American psychologist William James said that it was for the infant, "a blooming, buzzing confusion." We would have more impressions and information coming in than we could possibly cope with.

In the higher mammals the process of selection is to some extent under conscious control. The familiar "cocktail party effect" of blocking out all sounds except the voice of the person one is speaking to is one example of conscious selection. A mother who is deeply sleeping yet alert to the slightest sound of distress made by her baby is another. Man's nervous system, as distinct from that of the frog, can be programed. In chosen circumstances we can focus or increase our sensory receptivity for short periods of time, and not only for survival needs—for instance we have to sharpen our hearing when listening to music and our visual sense when looking at pictures. But this faculty of programability does not mean that we have our senses completely under conscious control. We certainly haven't, and, as Marais' experiments with the Boer girls showed, human beings have potential sensory acuity that, like a deeply-buried memory, is unknown to us in our normal state of consciousness, but may be awakened at other times, such as under hypnosis or in a situation of crisis.

The fact that this biological censor appears to be at work, limiting sensory inputs to the brain, has led some people to propose that man possesses a sixth sense, which under normal circumstances in the majority of people is completely suppressed by the censor. Professor Charles Richet, the French Nobel Laureate physiologist who died in 1935, coined the term "cryptesthesia" for a hidden sense—a means of perception using a mechanism which we know only by its effects. This sense includes all the functions generally called psychic, such as telepathy, clairvoyance, precognition, and dowsing. The question is whether this sixth sense really exists, or whether the perceptions attributed to it can be dismissed as delusions or explained in terms of the known senses, and it has produced a great deal of heated discussion over the last century. But as well as argument a great deal of research has been done, which has established beyond doubt that the human mind and senses have supernormal potential power.

The really difficult question is whether this potential is supernatural. Scientists reckon that the subject only really came into their province when it was separated from the realm of the supernatural, and they have no wish to see it dragged back by a few inconsistencies such as telepathy, clairvoyance, and precognition. If such things occur, they say, can we not explain them in terms of known science without calling upon supernatural explanations involving the existence of spirits, astral bodies

The Importance of the Frog's Eye

Experiments in Distorted Vision

separable from the physical body, or a sixth sense for which there is no detectable physical organ? This seems a reasonable question to ask. Surely we cannot begin to speak of extrasensory perceptions until we have fully understood the workings of the five known senses as well as their limitations.

The faculty man is most dependent upon, his sense of *vision*, is made up of two distinct parts. First, there is a *sensory* mechanism, in which a physical assembly of nerve cells functions to capture information. Second, there is a *perceptual* mechanism, a means of coding the information input. The sensory mechanism is in fact extremely defective, and we only see as efficiently as we do because the perceptual mechanism which is controlled by the brain corrects the eye's sensory errors. Seeing, in fact, is finally a function of the brain, which constructs images of the external world out of the bits of information conveyed to it by the nerve cells of the eye.

The brain's versatility in structuring visual information was demonstrated by Dr. Anton Hajos at Innsbruck University in Austria in the early 1960s. For several weeks Dr. Hajos and some of his students wore goggles equipped with prismatic lenses,

Professor Ivo Kohler performed further experiments using Dr. Hajos' prismatic goggles to investigate the nature of vision. Above: a crayon seen through a pair of spectacles. The lenses are half distorting and half plain glass.

Right: which is the right direction to walk?

which distorted angles, made straight lines appear curved, fringed sharp outlines with color, displaced objects in the wearer's field of vision, and made things seem to jump about whenever he moved. Obviously the wearers of these distorting goggles had to move about with the greatest care at first, but gradually the distortions and peculiarities reduced and after about six days were gone completely. One student was even able to ride a bicycle in traffic wearing the goggles. His brain had learned to construct a workable picture of the environment out of all the chaotic and conflicting information fed to it by the optic nerve.

After their brains had learned to make these complex adjustments, Dr. Hajos and his students found that when they removed the goggles the world seemed confusingly distorted. It took several days to readjust to seeing normally. A colleague of Hajos' at Innsbruck, Professor Ivo Kohler, then discovered another odd phenomenon. When subjects wearing the prismatic goggles went into a room lit by pure sodium light—which is made up of only yellow light, with no trace of other colors in it—they did not see everything in yellow, as other people did, but instead objects appeared to them in a whole range of glorious colors. The colors were not in fact present, but were created by the brains of the goggle-wearers.

How we tell the difference between colors, and to what extent other animals distinguish them, remained a mystery until the 1960s. In 1963 four research groups working independently— one in Germany, one in England, and two in the United States— discovered that the light-sensitive layer of the *retina*—a multi-

Left: a building seen through wedge prisms. It makes the environment seem to be a rubber world.

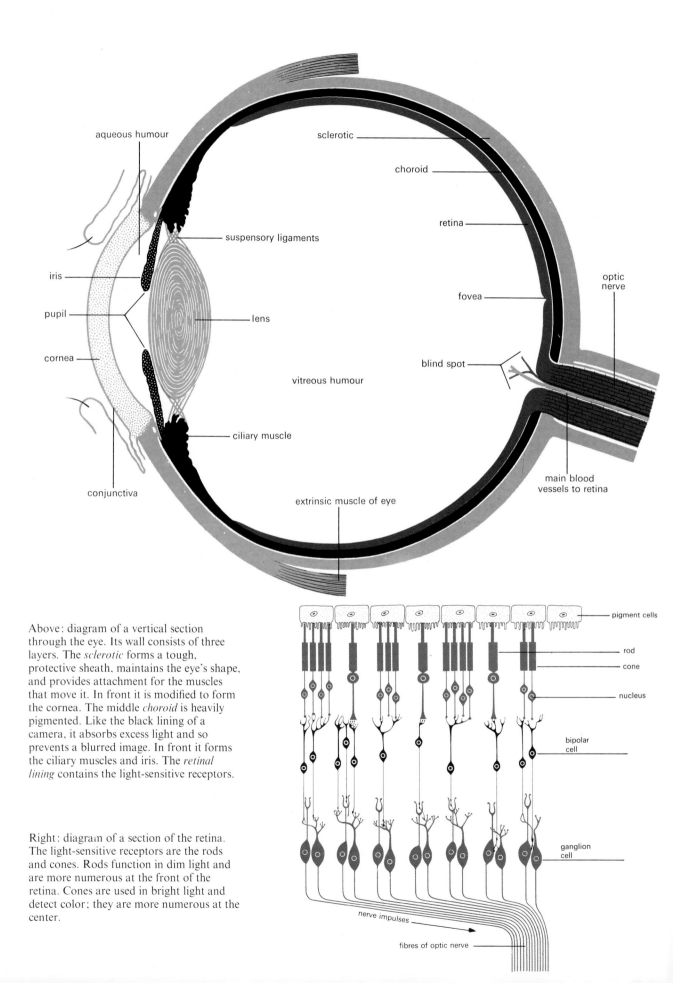

aqueous humour

sclerotic

choroid

suspensory ligaments

retina

iris

fovea

optic nerve

pupil

lens

cornea

blind spot

vitreous humour

conjunctiva

ciliary muscle

main blood vessels to retina

extrinsic muscle of eye

pigment cells

rod

cone

nucleus

bipolar cell

ganglion cell

nerve impulses

fibres of optic nerve

Above: diagram of a vertical section through the eye. Its wall consists of three layers. The *sclerotic* forms a tough, protective sheath, maintains the eye's shape, and provides attachment for the muscles that move it. In front it is modified to form the cornea. The middle *choroid* is heavily pigmented. Like the black lining of a camera, it absorbs excess light and so prevents a blurred image. In front it forms the ciliary muscles and iris. The *retinal lining* contains the light-sensitive receptors.

Right: diagram of a section of the retina. The light-sensitive receptors are the rods and cones. Rods function in dim light and are more numerous at the front of the retina. Cones are used in bright light and detect color; they are more numerous at the center.

The World in the Eye of a Bee

Bees see three main colors—yellow, blue, and the short ultraviolet wavelengths which are invisible to humans. Many flowers which seem to be a single color can be seen, when photographed under ultraviolet light, to have a distinct ultraviolet pattern.

Left: honeybee lighting on a seemingly dull plant.

Below: an evening primrose flower photographed first in normal light then in ultraviolet light.

layered coating at the back of the eyeball—consists of millions of receptor cells which fall into two distinct categories. They are known as *rods* or *cones*. The researchers found that rods detect only shades of gray, but that cones are color receptors of three types, each type responsive to a different section of the color spectrum. Various species of animals and birds have different proportions of rods and cones. The distribution is related to the conditions under which the animals live; for instance, nocturnal creatures have a predominance of rods.

If an alligator is made to choose over and over again between two cardboard disks, one colored and the other gray, and he is fed only when he chooses the colored one, he will eventually become neurotic. He is incapable of making the distinction between the two, and the sense of insecurity brought on by the randomness of his food supply will cause him to lapse into neurotic melancholia. He may crawl into a dark corner and lie there motionless for days. Most other animals also have poor color discrimination, and it is very doubtful whether any bull ever "saw red." We know that the frog sees blue and green because it will be repelled by green and jump onto anything blue. This is of course an illustration of the wisdom of instincts, because in dangerous circumstances the frog's safest place is a pond and not on grass. The primates are better endowed with color vision than most other animals, although the bee is exceptional in that it can see ultraviolet light. "Anyone who could look at the world through a bee's eyes," wrote the 20th-century Austrian zoologist Karl von Frisch in his book *The Dancing Bees*, "would be surprised to discover more than twice as many kinds of bloom as our ultraviolet-blind eye can see, with ornaments never registered before."

Bats and Dolphins

The electromagnetic spectrum consists of wavelengths ranging from less than one billionth of a meter to over 1000 meters, but only a tiny section of it, between 400 and 700 billionths of a meter, is visible to the human eye. This "visible spectrum" shades at one extreme into ultraviolet and at the other into infrared, which are light waves for which other organisms have evolved specialized active receptors. But the bee, although it can see ultraviolet light, is blind to red. For all his limitations, man possesses the widest ranging faculty of color discrimination.

Sharpness and distance of vision are dependent on the density of cells in a small pit at the back of the retina known as the *fovea centralis*, which is extremely small. Numerous rapid eye movements are required to constitute a large and detailed image. The density of cells thins out at the edge of the fovea centralis, which is why objects glanced at sideways look blurred. Elephants and rhinoceroses have as few cells in the fovea centralis as man has at the edge of his retina, so their world must look permanently blurred. On the other hand, the hawk is equipped with a cell density giving it vision equal to that of a man looking through binoculars that magnify objects eight times.

The next most valued sense in man is his *hearing*. Most human communication is conducted within a frequency "speech band" of between 300 and 3500 cycles per second, but many creatures emit sounds beyond this range. Recent research with whales and dolphins, for instance, has shown that these aquatic mammals probably have a highly developed language, although most of their communication is inaudible to us because their speech band is between 500 and 100,000 cycles per second. But if the human ear were attuned to the high frequencies of ultrasonics, then the silence of the countryside, for which many city-dwellers yearn, would not exist, particularly at night when multitudes of bats wage ceaseless ultrasonic war with their prey.

Both bats and dolphins use sound to explore their environ-

Below: bats in flight photographed in Tamana Cave, Trinidad. Bats fly and locate their prey through the use of high-frequency sounds. They emit a series of clicking noises which are usually inaudible to the human ear; these sounds bounce off objects and can be "read" as echoes. They "hear" on the same basis as the sonar systems used in submarine navigation.

ment and locate prey. They emit high frequency sounds in rapid sequence and "read" the echoes that come back when the sounds bounce off objects. These high frequency sounds can be beamed like a searchlight—a bat can locate a moth at a distance of 20 feet and keep it "on beam" by increasing its emission rate of sounds as it bears down upon the moth. A rate of about 12 clicking sounds a second is the bat's normal transmission in flight, but when approaching its prey it can produce an incredible 300 clicks a second. Ultrasonic recorders have measured bats' screams as reaching a volume of 100 decibels, and when we consider that a pneumatic drill registers 90 decibels we can imagine what the "peaceful" countryside would sound like at night if we could hear this range of sound.

A dolphin has a bundle of some 125,000 nerve fibers linking each ear to the cortex of its brain, whereas man has only about 50,000. Dr. John Lilly, the pioneer American investigator of dolphin intelligence, has estimated that whereas man processes *visual* information 10 times more efficiently than the dolphin does, the dolphin is 20 times more acute and efficient *acoustically* than man is. Lilly also points out in his book *The Mind of the*

Above: bottlenosed dolphin or porpoise (*Tursiops truncatus*). It has been known since Greek and Roman times that the smaller whales and dolphins (the cetaceans) are highly intelligent animals, and various mutually helpful arrangements between them and humans for catching fish have been described by fishermen from southern France to South America and Australia. More recently we have become aware that dolphins actually "speak" to each other. Certainly they use ultrasonics to "hear" in the same way that bats do.

Right: the American neurologist Dr. John Lilley listens to the sounds coming from a dolphin's blowhole. He claims dolphins speak a language of their own and even hopes to communicate with them eventually. A dolphin's brain is primarily adapted for the transmission and reception of sound, much more so than the human's is for all of our other senses put together. For a human to enter the dolphin's "world of sound" would be like a blind man opening his eyes —only much more dramatic.

Below: diagram of the human senses of smell in the nose and taste on the tongue. The two senses are closely related, each sensation depending on soluble compounds touching a sensitive surface. In general, the sense of smell helps animals to locate objects whereas the sense of taste helps them to judge what is being eaten.

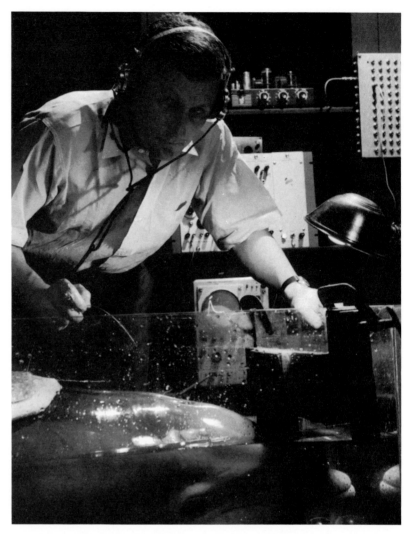

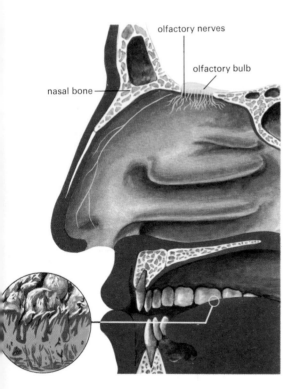

olfactory nerves

olfactory bulb

nasal bone

Dolphin, published in 1961, that since sound waves in water can penetrate a body without being reflected from or absorbed by the skin, muscle, or fat, reflecting only from air cavities and bones, so dolphins must be able to "see" into each other's internal organs. He suggests that the love-talk of dolphins might go something like this: "Darling, you do have the cutest way of twitching your sinuses when you say you love me. I love the shape of your vestibular sacs." This may be pure fancy, but it is a fact that ultrasonics can be used to "see" inside the body, and this method has recently been put to advantage in medical technology. Today ultrasonic scanners are proving a much safer tool than X-rays for medical checking and diagnosis, particularly during pregnancy.

Another diagnostic tool designed to develop and increase human sense perception is the "artificial nose" invented by Dr. Andrew Dravnick at the Illinois Institute of Technology in the United States in 1965. As every disease produces characteristic body secretions, Dr. Dravnick proposed that his device might be used to give early warning of developing illnesses. The "artificial nose" works by blowing a current of pure air over a body in a confined space, then sucking off the air and submitting it to automatic chemical analysis. The machine breaks down the human smell into 24 parts and can create a "scent register" of an individual, based on the way his various body smells are combined and distributed. This register can then be stored away and used to identify him on other occasions, just like a fingerprint. Needless to say, police forces have expressed an interest in Dr. Dravnick's invention.

Police the world over have, of course, been using dogs to track and identify people for quite a long time. Whereas man has about five million sensory cells in the *olfactory* or "smell-sensitive" area of his nose, some breeds of dog have over 200 million. But these figures do not give a true idea of the difference in sensitivity between the two, for it has been estimated that a dog's sense of smell is a million times better than a man's. When a dog tracks a man it picks up the scent of sweat molecules forced out through the soles of his shoes every time he takes a step. Millions of such molecules are left behind with each step, and for a dog they produce a highly individual smell. In one of many experiments with tracker dogs, 11 men walked in single file, with the dog's master in the lead and the others following in his footsteps. After a time the file divided into two. When the dog

Smell-sensitive Nasal Areas

Below: the acute sense of smell of a Labrador dog. Here it follows a trail through the snow. It has been estimated that a dog's sense of smell is a million times better than a man's.

Above: the angle shades moth. The moth's acute sense of smell is shown in its highly developed antennae.

reached the dividing point it had no difficulty determining the file that was still led by his master.

The human nose is believed to contain about 14 different types of receptor cell, compared with the dog's total of 30 or more. If each type of cell is receptive to a different scent, different combinations of the 14 or so primary odors will constitute a particular smell. This registers in the brain through impulses which are triggered in the receptor cells by each of the individual scents. It has been estimated that in man eight molecules are required to trigger an impulse in a nerve ending, and that for a smell to be detected approximately 40 nerve endings must be triggered. In dogs and many other creatures smells do not need to be so strong for detection. Some male moths and butterflies can locate a female from miles away if just one or two molecules of the "love scent" she emits are carried to them on the wind.

Moths and Pigs

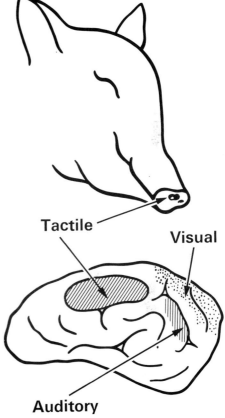

Left: perhaps surprisingly, the pig has an acutely well developed sense of taste, much more sensitive than man's.

Tactile

Visual

Auditory

Above: the "porcunculus." This drawing, done in 1943 by the British physiologist Lord Adrian, shows the huge area in the pig's somatic-sensory cortex that is devoted to the touch and taste faculties of the opposite half of its snout. This patch alone is bigger than its entire auditory cortex and as large as its visual cortex.

Taste is related to smell in that it is also based on a sensory response to a chemical stimulus, but it is a much less discriminating sense. For a taste to register requires about 25,000 times as many molecules as are needed for the detection of a scent. There are only four primary tastes: sweet, sour, bitter, and salty, and each taste bud is responsive to only one of these, depending on where it is located on the tongue. The buds on the tip of the tongue register sweetness or saltiness, those on the side register sourness, and bitterness is registered by the buds at the back of the tongue. In all, man only has about 3000 taste buds, whereas for example the pig has double that number. Although connoisseurs of wine and food make much of the delights of taste, the pleasurable sensations they experience are in fact derived from the scents which rise from the cavity of the mouth and are registered by the olfactory cells in the nose. People suffering from heavy head colds are unable to taste anything, and the most delicious wine will seem tasteless if you try drinking it while holding your nose.

Pressure, pain, heat, and cold are different aspects of the sense of *touch*, and it is believed that distributed over the surface of the skin are specialized receptor cells for each of these sensations. The cells are obviously more densely clustered in certain areas, particularly the so-called erogenous zones of the body that are highly sensitive to sexual stimulation.

Many animals possess a sensitivity to vibration which is related to the sense of touch, and the phenomenon of animals showing alarm or panic some time before an earth tremor occurs has been noticed all over the world. This observation was scientifically measured in 1960 when a German zoologist, Dr. Ernst Kilian, was in Valdivia in southern Chile. As it happened, a series of large and small earthquakes hit the town over a period of days. Showing admirable scientific dedication under such circumstances, Dr. Kilian kept a number of different animals under observation and systematically recorded the intervals that elapsed between their manifestations of panic and the occurrence of a tremor.

From time to time the phenomenon of "skin vision" has been noted. In the 1920s the French writer Jules Romains published a book in which he reported numerous tests he had conducted. These had convinced him that many people possessed or could develop what he called "paraoptic ability"—that is, the ability to "see" by touch. In the 1960s a Russian girl named Rosa Kuleshova astonished scientists by her apparent ability to read a book just by running her fingers over the print. Widespread investigations turned up a number of seemingly similarly talented subjects throughout the Iron Curtain countries, but many were found to be cheating, and the phenomenon was discredited. However, in Los Angeles, California, a woman named Mary Wimberly who had been blind since birth heard about the Russian investigations and offered herself as an experimental subject. The parapsychologist Dr. Thelma Moss tested her over several months, and their experiments produced some interesting results.

In their first session together, Dr. Moss gave Mary the task of sorting sheets of completely black or white paper into their

Right: Mary Wimberly and her guide dog. After more than a year of trial-and-error experimentation a technique was evolved through which Mary would try to develop "skin vision" or eyeless sight. Russian researchers claim that about 30 percent of the normal population can be taught skin vision. Although she has been quite successful in laboratory tests, Mary still does not know consciously how she "sees" colors, and she personally will not feel a sense of achievement until she does know, in order to pass on the knowledge to other blind people.

respective piles. At first the results were disappointing, no better than could be expected by chance. Mary returned to Dr. Moss' laboratory daily, however, and after three weeks she had learned to separate the black and white sheets with uncanny accuracy. But when Moss added a third color, red, the results dropped sharply to chance level again. Persistence was rewarded, however, when after further practice Mary was able to distinguish by touch six colors and to sort them into piles, even working in total darkness. When a rigorously controlled experiment consisting of 1500 trials was mounted and the results were assessed by a statistician, Mary's success rate above chance expectancy was five million to one.

One thing that stands out very obviously when we consider the five sensory functions of humans and animals is that each of the human senses covers a limited band of a spectrum. Many phenomena often attributed to a sixth sense can be explained in terms of supernormal functioning of one of the primary senses. Some aspects of animal behavior seem little less than miraculous or supernatural to us, but in fact the animal is behaving entirely naturally by hearing, seeing, smelling, or feeling something beyond the limits of normal human sensitivity. Eugene Marais' experiments in hypnotic hyperesthesia demonstrated that those limits can be transcended; but on the whole the limits on individuals of the human species, as on those of other species, are so consistent that they can be considered "species specific" characteristics.

If many apparent sixth sense phenomena can be explained in terms of one of the five senses, this does not necessarily mean that there are only five senses. In fact animal studies have shown conclusively that there are more. Some birds have a magnetic sense, so that when migrating they navigate according to the earth's magnetic field. Snails have a sensitivity to ionizing radiation, and they will retract their horns in its presence. Paramecium will swim along an electric current. These sensitivities to magnetism, radiation, and electricity are alien to man, but they are nonetheless quite natural. It is conceivable that some people may possess them to a certain degree, and some investigations of ESP take them into account.

But when these additional animal senses have been taken into account, as well as the possibility that one of the five normal senses may be hyperacute, the mystery of man's so-called psychic faculties remains. There are numerous cases on record of people suddenly becoming aware of events occurring at a distance, particularly in circumstances when a loved one dies or is in danger. In laboratory experiments over recent years modern technology has made possible demonstration of the interaction of minds, unaided by communication through any of the known sensory channels.

The development of physiological recording devices has enabled parapsychologists to show that telepathy probably goes on between people all the time. In a famous experiment, the researcher E. Douglas Dean of the Newark College of Engineering in New Jersey wired a subject to a plethysmograph, a device which measures fluctuations of blood volume in the fingertips. In another room a person was given a list of names and asked to

Experimenting With the Blind

concentrate on one of them at a time in random order. When he concentrated on the name of a person emotionally connected with the experimental subject, the plethysmograph registered significant changes in the subject's blood volume, but it did not register any changes when the "sender" thought about an unfamiliar name.

In a similar experiment conducted by Charles Tart at the University of California at Davis, a subject was wired up to physiological recording devices. He was then required to guess when electric shocks were administered to a second subject in another room. The second subject was given shocks at random intervals, and at these times significant physiological changes were registered in the first subject. His conscious guesses, however, were not particularly accurate.

In a third experiment of this type the physicists Harold Puthoff

A Biological Censorship?

Left: some birds appear to have a "sixth sense" which makes them sensitive to magnetism. This may be how migratory species, like these snow geese at a wildlife refuge in South Dakota, are able to navigate so well even when the sun is hidden.

and Russell Targ of the Stanford Institute in California used flashes of stroboscopic light instead of electric shocks and monitored the two subjects' brainwave activity. They confirmed Tart's finding that, although the second subject was unable to consciously guess when the other was subjected to a stimulus, he seemed to "know" it at a physiological level.

These experiments and many others over recent years appear to demonstrate that human beings possess senses of which they are not normally aware, even when those senses are functioning. Such investigation also appears to bear out the idea of the operation of a kind of biological censorship of the psychic functions. Whether that censorship is biological wisdom or folly—something that facilitates man's highest development or inhibits it— is another hotly debated mystery, and one that each of us must ponder for himself.

Chapter 9
The Tides of Living

Mysterious body rhythms, different yet similar for each individual, affect most aspects of our lives, sometimes without our being aware of them. What are they, how do they work, and how can we use them to function better? These rhythms also affect animals and plants, and it is from observations by naturalists that we first came to know of their existence. The usefulness of drugs, efficiency of surgery, and the cure rate of cancer may all be increased by taking biorhythms into account. Humans are also affected in mysterious ways by light and darkness, particularly in determining cycles of sleep and waking. What happens when we sleep? We are only now, with our growing awareness of the tides of life, beginning to understand the truth of the old Arab saying about the speed at which the soul travels.

One thing the South African naturalist Eugene Marais observed that baboons have in common with man is a tendency to what he called "Hesperian depression" or "evening melancholy." Every evening at sunset in the native villages of the African veld, an air of quietness and dejection would fall. All conversation, laughter, and games would cease, and the old would gather in silent groups while the young crept close to their mothers. Then after sunset, fires would be lit and conversation, activity, song, and laughter would be heard again. Likewise in the baboon troop, the animals would cluster together at sunset, the old in attitudes of profound dejection, the infants whimpering in their mothers' arms. To compare the two communities, Marais wrote, was "to realize beyond any shadow of doubt that you have here a representation of the same inherent pain of consciousness at the height of its diurnal rhythm."

The fact is that human beings go through daily rhythms of change in mood, abilities, wakefulness, and biological functioning. Some of these changes are linked with environmental influences and are shown by other animals. The conditions of human life in cities, however, have hidden these rhythms and, until recent years, psychologists, physicians, and employers have ignored their existence. Men in cities and industrial societies have lived as if they were machines rather than biological organisms, expecting to perform their tasks with equal efficiency at all times. We have recognized no biological rhythms except the most

Opposite: sunset over the sea off Maui Island, Hawaii. "Evening melancholy" afflicts both baboons and man—the twilight hour seems to hold the threat of an unknown future or end of a familiar routine. This feeling may be related to the daily rhythms which regulate the lives of the higher primates.

obvious ones of the menstruation cycle of the female and the daily alternations of sleep and waking. One of the aims of rational man—one that has been realized in that most remorselessly rational of man's enterprises, industrial production—has been to banish sleep. Many people consider it a weakness to need to spend a third of their life asleep. But in the last quarter century many things have been discovered about sleep, as well as many other biological rhythms, that suggest inevitable failure for the man who would like to be as consistent and efficient as a machine. In fact modern investigations of man's internal rhythms, and of how these processes are affected by environment and way of life, have helped us to an impressive degree of self-knowledge.

Marais wrote of the "diurnal rhythm" of life, employing a term not much used today. Traditionally "diurnal" suggests the opposite of nocturnal, implying that the rhythmic flow of life ceases with the coming of night and sleep. Today the preferred term is "circadian rhythm," coined by Dr. Franz Holberg of the University of Minnesota Medical School in the United States in 1959, because it indicates that the internal tides of life, the *biorhythms*, ebb and flow in a 24-hour cycle. In about the same time as the earth takes to turn on its axis, numerous processes in the human body go through cyclical changes: temperature; blood pressure; rate of respiration and of cell division; levels of sugar, hemoglobin, and amino acids in the blood; production of urine, of enzymes, and of adrenal hormones; all are on a program lasting approximately 24 hours.

Obviously our 24-hour cycles of life are mainly determined by social factors, by our schedules of work, rest, and recreation, and by the fact that our non-body regulators, our watches and clocks, are designed for convenience to measure time in terms of 24 units. If clock time and biological time coincided exactly, it would suggest that the human body might be adaptable to arbitrary schedules, but in fact this is not the case. Several studies have been made of people who spent long periods, sometimes months, in isolation in caves or special rooms, shielded from

Biorhythms

Opposite: production line in a citrus fruit factory in Cyprus. Mass production demands machinelike efficiency from humans in order to keep the line moving, sometimes 24 hours a day. But man does not perform at one level throughout the day, nor perhaps should he attempt it.

Left: stationary activity cage (left) and kymograph used for recording short periods of activity in rats. Every movement of the rat, however slight, is transmitted pneumatically to the smoked drum of the kymograph, which measures periods of $1\frac{1}{2}$ to 2 hours. The purpose of the experiments was to measure the rats' circadian rhythm, and they were performed by Curt Paul Richter, a professor of psychobiology at Johns Hopkins University Medical School in Baltimore, Maryland.

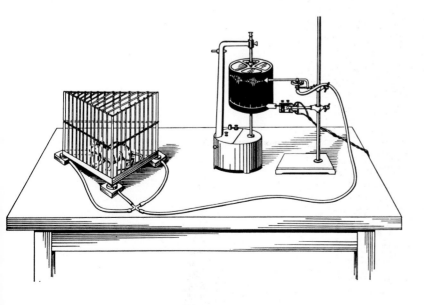

Right: Nathaniel Kleitman and B. H. Richardson emerge from Mammoth Cave, Kentucky in 1938 after 32 days in a damp and chilly subterranean chamber. During this time they lived in near total isolation from the outside world, attempting to adjust to a 28-hour daily schedule of 19 wakeful hours and 9 hours asleep. During the experiment the two subjects regularly sampled their own body temperatures and carefully evaluated their sleep.

external influences and clues to the passing of time. Invariably, after an initial period of adjustment, they settled down to individual physiological cycles, generally lasting rather longer than 24 hours. Two French volunteers, who lived in separate underground caves for three months in 1964 and kept records of their physiological processes, found that they lived in 24.6-hour and 24.8-hour "days" respectively. Statistical analysis of a number of such experiments produced a figure of 25.05 hours as the average cycle. Psychological tests established that people who "cycled" at approximately this amount of time were more stable than people whose circadian cycles deviated significantly from it.

Researchers at the Max-Planck Institute in Munich came up with a suggestive finding during a research program in which certain conditions were systematically varied. They shielded one of two experimental rooms against magnetic and electric fields, and found that people who lived for a time in the shielded room lived on average a 25.26-hour "day," whereas those who lived in the unshielded room had an average cycle of 24.84 hours. The difference implied that human biorhythms are affected even by the weak electromagnetic field of the earth, which further implied that man is much more intimately related to his environment than he generally acknowledges.

Thus it appears that man's natural physiological circadian cycle is about 25 hours, and that under normal circumstances he compresses it to 24 hours for social convenience. The question whether he can compress or extend his cycle by an even greater margin has been investigated by a number of researchers, including those working with the American and Soviet space agencies. Soviet cosmonauts in orbit have been kept on a normal 24-hour activity/rest schedule, but their American counterparts have been tested on various schedules, such as alternating four-hour periods of rest and activity, and have afterwards complained of lack of rest. The Soviets base their policy on the results of a prolonged and detailed study of volunteers who lived 18-hour days in isolation chambers, and who then showed increased likelihood of infection, impaired coordination, and physiological signs of stress, such as excessive ascorbic acid in their urine. In a French experiment three men tried to live a 48-hour day, working for about 36 hours and sleeping for about 12. Apparently they were able to do so for months on end, but two of them later confessed that they had felt drowsy all the time, and the temperature cycle of one of them was found to be a normal 24 hours and 44 minutes. The upshot of all such studies has been the finding that although a person may force himself to live a longer or shorter day than the normal, his physiology will not adapt and his biological processes may fail to mesh with each other, resulting in generally impaired functioning.

Every healthy body has a multitude of internal "clocks" which time and control different processes so that they work together and harmonize like a symphony orchestra. As the American science writer Gay Gaer Luce has written in *Body Time*, her summary of research into this subject published in 1972: "We

Physiological Life Cycles

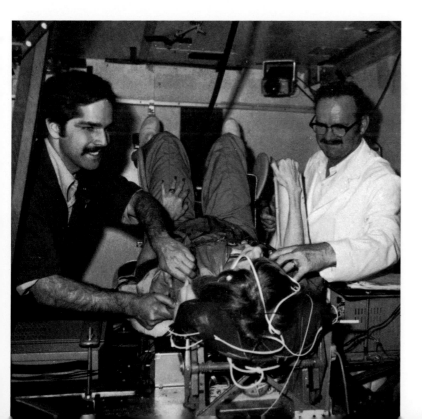

Left: Dan Gundo (left) and Sil Corpus, centrifuge technicians at Ames Research Center, assisting a female volunteer into position on a 20-G centrifuge. She was taking part in the preliminary stages of a Space Shuttle Bedrest Study conducted by the U.S. National Aeronautics and Space Administration (NASA), designed to investigate the sleep cycle requirements of astronauts.

"Jet Lag" and Other Problems

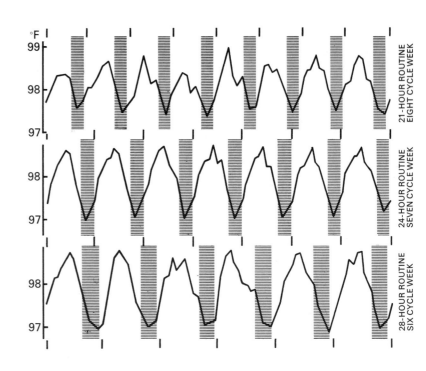

Right: weekly body-temperature curves of a subject under three different routines of sleep and wakefulness. Upper curve: 21-hour cycle of 15 hours awake and 6 hours asleep; middle curve: the usual 24-hour cycle of 17 hours awake and 7 hours asleep; lower curve: 28-hour cycle of 19 hours awake and 9 hours asleep. The shaded areas represent time spent in bed, usually in sleep. Each curve is based on five weeks of following one or another routine. Eight, seven, and six well-defined body-temperature waves can be seen, depending on the number of cycles lived per week, and the lowest temperatures always occur during sleep. The studies were carried out under the direction of the American scientist Nathaniel Kleitman, one of the most respected authorities on sleep and wakefulness.

Below: Greek relief of the 5th century B.C. showing an offering to Aesculapius, the Greek god of medicine, and Hygeia, goddess of health. The Greeks applied therapy in cycles, in which patients were given treatment or exercise in rotations of three or seven days. Periods of seven were considered vitally important, perhaps underlying the seven-day week.

now know that we are transformed, hour by hour, as our nervous systems, metabolism, and vital organs fluctuate in circadian rhythms. Thus, it is not surprising that our abilities, our vulnerability to harassment, trauma, or infection, the acuity of our senses and our symptoms of disease all vary, as well, in circadian fashion." She continues, "It is no longer surprising that the drugs we take affect us differently depending upon the time we take them. This same circadian rhythmicity has implications for the way we learn, and what memories influence our lives." In these few short sentences Gay Luce summarizes the findings of a host of researchers, findings which as yet few psychologists, educationalists, physicians, and pharmacologists take seriously. There is today a developing movement towards *holistic medicine*

—that is, the treatment of an organism as a whole rather than of symptoms affecting only a part of it—and although the approach is hardly new (it was recommended by Hippocrates, the Greek "Father of Medicine," some 2400 years ago), the implications of natural rhythms in body processes have yet to be incorporated into general medical practice.

Gross disruptions of body rhythms are likely to be caused by rapid travel across the earth's time zones—"jet lag" is one widely acknowledged phenomenon in which time and the body are seen to be interactive, although most people see it simply in terms of adapting to the social activity/sleep schedule of the new location. Airlines, naturally concerned that their flight crews maintain maximum efficiency, have carefully studied the effects of jet lag. They try to plan their crews' work/rest schedules so as to minimize the ill effects of frequent rapid travel across time zones. In one study, after a flight from the United States to Rome, a flight crew performed below average on psychological tests, and their body temperature and heart-rate cycles took six and eight days respectively to adapt to the change. Airlines urge long-haul pilots to remain as close to their home schedule and

Below: jet lag involves a shift in phase of the circadian cycle. Traveling on an airplane from the east coast of the United States to Europe and back would require a process of desynchronizing and then resynchronizing the body's internal time. The outer area of the squares shows Greenwich (universal) Time. The outer large circle gives local geographic time and the smaller inner circle the physiological time of the traveler. The black section of the outer circle indicates night and the shaded section of the inner circle the normal time of sleep.

LOCAL TIME: MIDDLE EUROPE   CYCLE DESYNCHRONIZATION

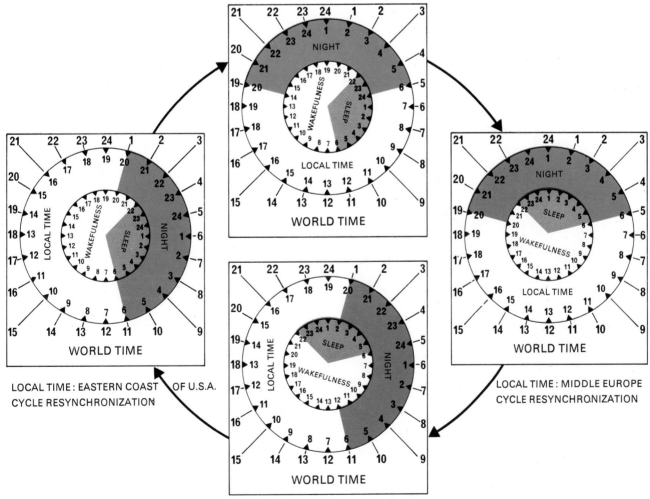

Drinking Times

habits as possible in order to avoid upsetting their body rhythms. Jet-traveling political and business negotiators have also in recent years become aware of the importance of rest and timing of crucial meetings after a journey if they are to function at their best. The old Arab saying that the soul travels at the speed of a trotting camel may not be scientifically precise, but it holds an insight into the rhythmic basis of life that our science is just getting around to understanding.

Although the immediate effects of physiological disruptions caused by travel can be minimized, the cumulative long-term effects are still unknown. Mice subjected once a week to a reversal of their lighting schedule, equivalent to a jet flight halfway around the world, have a 6 percent shorter life span than mice living in a constant light/dark cycle. Of course men are not mice, but in physiology and metabolism the two species are very similar, and the possibility that decreased life expectancy may be an occupational hazard for long-haul pilots cannot be lightly dismissed. Nor can the possible long-term effects of habitual taking of drugs or medications. Experiments with animals have shown that some drugs may have no appreciable effect on body cycles while they are being taken, but that when treatment ceases the rhythms may go seriously amiss.

Medicine as practiced in Europe and America is predominantly *allopathic*, that is, based on treatment by drugs, so physicians ought to be more concerned than they generally are with the timing of interactions between prescribed drugs and natural body chemistry. Vulnerability to the effects of drugs may vary greatly throughout the day with the body's rhythms of peaks and lows in the production of different substances. Many people will be stupefied by drinking a little alcohol in the middle of the day, but they may be able to drink five or six times that amount in the evening. The same dose of amphetamine administered to groups of mice at different times of the day may kill as many as 77 percent or as few as 6 percent of them. At a hospital for asthmatic children, hormone medications were systematically given to different groups at different times, and it was found that the children who were given theirs at 1 a.m. or 7 a.m. benefited more than the others. These facts indicate that the body does not react to drugs in the same way at all times. Desired results may be obtained by giving small doses at the right times, whereas larger doses given at other times would be both less effective and likely to produce unwanted side effects. It is debatable whether medicine should be as drug-oriented as it is, but doctors should certainly take into account any possible techniques for lowering drug dosages and increasing their effectiveness. The practical difficulty, of course, is that everyone's physiological circadian time chart is different, and that in order to prescribe a schedule of medication the physician would need to have the patient under continuous observation for at least 24 hours. But technology has a way of overcoming difficulties when the need is sufficiently urgent—possibly in this case miniature monitoring devices could be produced and implanted in or worn by the patient.

Surgery, too, could benefit from knowledge of a patient's physiological time chart. A person's pain threshold, his suscep-

Below: the effects of drink depend upon the time of day it is taken, even in 16th-century Persia. Here two youths are having a good time in the evening.

tibility to anesthetics and vulnerability to infection, the reaction of his nervous system to the trauma of surgical incision, and his rate of natural tissue repair, all vary throughout the day. If surgery were patient-oriented, all these factors would be taken into account, and operations would be scheduled accordingly. But considering all the other factors involved, such as the time of the surgeon's highest competence, not to mention the availability of hospital facilities and staff, it must be rare indeed that surgery is carried out when it ideally should be. When transplant surgery is being performed yet another factor comes into play — the circadian rhythm of the living organ to be transplanted, and

Above: *Gin Lane*, engraving by the 18th-century British painter William Hogarth. Habitual day-long drinking does not carry the same connotation as an occasional glass in the evening.

Right: variation in hand-steadiness as related to time and body temperature. The graph expresses reciprocals of the ratios of test scores made in three 1-minute tests to the score at 8 a.m. (the control). Two subjects (initials N.R.C. and F.J.M.) were given five trials per day for 20 days. Upper curves: body temperature; lower curves: hand steadiness.

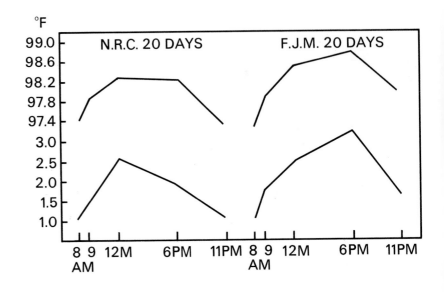

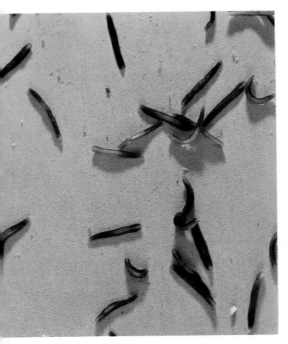

Above: marine flatworms (*Convoluta roscoffensis*). Flatworms migrate up to the surface of the sand when the tide ebbs so that green symbiotic algae can photosynthesize. The flatworms then burrow down into the sand just before the tide covers them. It is thought that the action of waves on the shore stimulates their responses, which show a circadian rhythm based on tide cycles.

whether it can be phased in with the rhythm of the recipient. Some surgeons believe that one cause of organ rejection in transplants may be that the body rhythms of donor and recipient did not match, as when two musicians play to different beats.

Diabetes is a disease caused by the failure of the pancreas to produce the hormone insulin, which metabolizes the blood sugars that the brain needs. In a normal healthy body, insulin levels rise and fall in the blood in a circadian rhythm, and they are "phased" with the production of other hormones. Excessive insulin dosage at the wrong time can disrupt other hormonal processes, so the successful treatment of diabetes depends on the right amount of hormone being given at the right time.

Cancer, too, is an illness in which the body's rhythms are grossly disrupted. There is a circadian rhythm of cell division, but cancer cells multiply independently and at a faster rate than other cells. Some researchers think that certain cancers may be produced by excessive and mistimed secretion of particular hormones, caused by stress or emotional trauma, and others have found that monitoring abnormal cell-division rhythms can indicate the existence of a cancer before any obvious physical signs have developed. In treatment of the disease by X-rays, too, the timing of the dosage has been found to be a factor in successful treatment.

How did the circadian rhythms of all living organisms, including the intricate intermeshing which occurs in complex organisms, originate? And what controls them? What or where are the triggers that set off each process as it is due?

The fact that plants and lower forms of animals have circadian rhythms indicates that such rhythms are not necessarily dependent on a brain and nervous system. "We are constructed out of time," writes Gay Luce, and the idea is today accepted by most biologists that time is a component of every living cell, that every cell has its inbuilt "clock." Evolution theory may provide the answer to the question of origins: in the process of natural selection, organisms with circadian rhythms would have a survival advantage in the conditions of life on a revolving earth. Rhythmicity must originally have been an inborn property of cells. As more complex organisms developed with specialized cells,

Inbuilt "Clocks"

Left: ostracods and copepods, planktonic animals which perform migrations up to the surface levels of the sea at night. At dawn they make their way down again to spend the daylight hours in deeper water near the bottom. This rhythmic migration pattern persists even when the animals are kept in conditions of constant light or darkness, when it is presumed that the pattern is maintained by a cycle of hormone secretions. These are controlled not by the light of day but by the animals' internal clock.

Above: many flowers show a rhythmic cycle of opening during the day and closing up at night. This chamomile flower was photographed early in the morning with its petals folded down and then later at midday (left) when the petals were fully spread. This is just one type of circadian cycle shown by plant forms.

Artificial Time Scales

so networks to trigger and synchronize physiological processes must have evolved. Some of these control centers have been located in the "old brain," the *hypothalamus*, as well as in the pituitary and pineal glands. A body functions most healthily and efficiently when these internal synchronizers operate in harmony with external synchronizers such as light and darkness.

Women with erratic menstrual cycles have been regularized by the simple therapy of leaving the light on in their bedrooms throughout the 14th, 15th, and 16th nights of their cycles. The same treatment has enabled some women who despaired of ever having children to become pregnant. Such therapy may sound like quackery or an old wives' tale, but it is based on the recently discovered fact that light acts as a synchronizer on the *endocrine* system of the body, the system that regulates the secretion and release of hormones into the lymphatic vessels and thence to the bloodstream. The constant light through the critical phase of the women's menstrual cycle triggered the release of the hormones that cause ovulation. This is one application of the new understanding of light's influence on life processes. It has long been observed all over the world that blind girls begin menstruation earlier than those with sight—but the phenomenon was a complete mystery until the relationship between light and hormonal regulation was discovered. People who go blind tend to have erratic hormonal rhythms, while, on the other hand, people who have successful operations for cataracts and are thus able to see again soon become stabilized in their circadian hormonal rhythms. Similarly, plants raised from seed in complete darkness

Below: an example of how light affects the natural circadian cycle of plants. Artificial lighting in a commercial greenhouse is kept on at night to lengthen the "days" and alter the usual ratio of light to darkness during the winter months. This stimulates stem growth of the young chrysanthemum cuttings by activating the spring/summer instructions of the plants' internal clocks.

show no rhythmicity, but if once exposed to light their natural circadian cycle is triggered into action, which illustrates the interaction of internal and external synchronizers. Light is the highest authority in the chain of command governing the plant's behavior, but detailed instructions for the circadian running come from inside. It seems that human beings function in the same way, that they have what computer technologists call "stored programs" in their body cells and endocrine glands, but the influence of light is needed to cue these programs into action.

The mysterious *pineal* gland—which has been called the "third

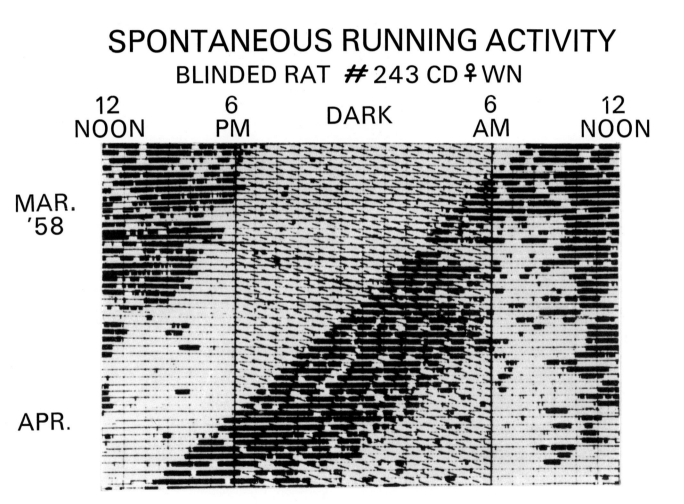

SPONTANEOUS RUNNING ACTIVITY
BLINDED RAT # 243 CD ♀ WN

| 12 NOON | 6 PM | DARK | 6 AM | 12 NOON |

MAR. '58

APR.

eye" and which the 17th-century French philosopher Descartes believed was the seat of the soul—is located in the midbrain. It is now thought to govern certain bodily functions, particularly sexual ones, and in turn to be influenced by light. Those rare children who are born without a pineal gland reach puberty extraordinarily early, sometimes as young as five or six, a phenomenon similar to the premature menstruation of the blind. The pineal gland secretes a unique chemical known as *melatonin* as well as a number of other substances that the brain and nervous system need. It is thought that melatonin production is influenced by light, and that this particular chemical is crucial in the timing of sexual development. The theory is supported by

Above: activity-distribution record for spontaneous running of a blinded rat, illustrating the amazing consistency of the times of onset of activity from day to day. The rat's circadian rhythm is directed by an internal time clock and is not dependent on the sun.

Above: detail of the creation of the heavens from the Cupola of Creation, a 12th-century Italian mosaic. The traditional Christian view according to Genesis is that God created the sun, moon, and stars and fixed them in the sky above the earth. It is only in this century that many scientists have become convinced of the major effect that light and the daily cycles of the sun and moon have upon all living creatures.

the finding that in rabbits raised in total darkness from infancy the development of the pineal gland is drastically retarded.

It is also theorized that long-duration biorhythms—such as the proverbial "spring fever" and production of the special hormone that helps humans to cool off in summer and some hairy mammals to grow and shed winter coats—are influenced by the changing proportion of light to darkness as the seasons progress. This theory, too, is supported by evidence collected in animal studies. In one experiment migrating birds were captured in autumn and divided into two groups. One group was kept in an artificial environment, in which the changing ratio of light to darkness was rapidly speeded up to simulate the passing of a whole year in six months. The other group was kept on a natural lighting schedule. By the next spring the birds that had experienced the accelerated year were physiologically ready for autumn, whereas the others were in their natural spring condition. When both groups were released in a planetarium in which the star patterns of the spring sky were represented, the autumnal birds flew south while the others flew north.

Of course the main effect that alternations of light and darkness have on the biorhythms of human beings is in determining their circadian cycles of sleep and waking. With the development

Biorhythms and Seasons

Long-duration biorhythms like "spring fever" are thought to be influenced by the changing proportion of light to darkness as the seasons progress.

Left: red deer stags in spring with the beginnings of their new antlers, still covered with their thick winter coats.

Left: red deer stags in summer, sporting a full-grown set of antlers and a short coat.

Above: *Hunters in the Snow* by the 16th-century Flemish painter Pieter Breughel the Elder, showing the activities and clothing appropriate to humans during the winter stage of the yearly seasonal cycle. Not only are environmental conditions different for each time of the year, but some internal human characteristics are also controlled by our internal clocks to adapt us to each season.

over recent decades of sensitive physiological recording devices, it has become possible for scientists to study what goes on during our nightly trips into oblivion. They have found that during a good night's sleep a person normally goes through four or five distinct cycles, each cycle consisting of four different stages of sleep which altogether last on average 90 minutes.

In the first stage of sleep the brain gives off electrical impulses at the rate of 9 to 13 cycles per second. These are known as alpha waves and indicate relaxation. In the second and third stages the waves become longer and slower, and 15 to 20 minutes after falling asleep a person enters the deep sleep of stage four. Here he remains until nearly the end of the cycle, when he goes through stages three and two again before entering what is known as REM sleep. The letters stand for "rapid eye movements," and during this stage of sleep a person's eyes will dart about under his eyelids. Researchers have found that it is during this period that we dream.

This process of descending into stage four sleep and surfacing into REM sleep is repeated throughout the night, though in the later cycles comparatively less time is spent at stage four and more in lighter sleep and the dream state. Modern sleep research has shown that dreaming is essential to man's psychological balance and well-being. Experimental subjects who are awakened whenever they enter REM sleep soon show symptoms of anxiety, imbalance, and irritability, and people who have been kept awake for days spend a greater proportion of time than usual in the dream state when finally they are allowed to sleep.

Deprivation of stage four sleep is both a cause and effect of extreme depression, stress, and anxiety. People incapable of sleeping deeply are often listless and lacking in vitality. Recent research suggests that this may be because in stage four sleep the hormone known as HGH (for Human Growth Hormone) is released into the blood. By taking blood samples from sleeping subjects at intervals throughout the night, researchers have found that hormones are spurted into the blood at different times and intervals according to a fixed program. In a normal healthy

Man and Seasons

Below: *Haymakers*, one in the series of six paintings illustrating the months and seasons of the year by Pieter Breughel the Elder.

Sleep and REM

Opposite: *The Magic Apple Tree* by the 19th-century British artist Samuel Palmer. Mystics and Romantic poets have always been convinced that there is a harmony between man and nature. Now scientists are agreeing with them to some extent, based on numerous experiments which show essential links connecting our internal cycles with those of the world around us.

person hormone levels are high when he awakens in the morning, and body temperature, after sharply declining in the middle of the night, has climbed back up to normal.

The approximately 90-minute cycles that affect sleepers are continued throughout our waking hours, too. As in sleep we drift from one state of consciousness to another, each having its physiological purpose, so also during the day we pass through alternating phases of concentration and mind-wandering, of vitality and relaxation, which come in cycles of about 90 minutes. This period has consequently been called "the biological hour." This daytime rhythm tends to be hidden by the requirements of

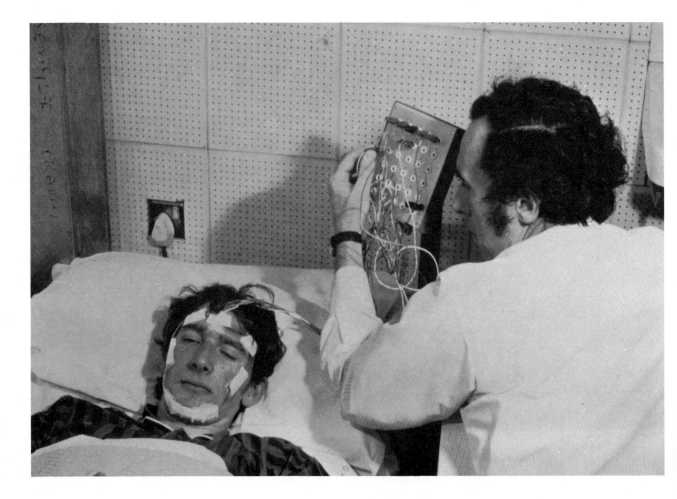

Above: Dr. Ian Oswald of Edinburgh University and a volunteer for a sleep research experiment. Small silver disk electrodes are attached to the subject's head and face, and wires lead from these into a head box. Electrical impulses pass from the head through a cable into a machine which records EEGs, thus giving a graph picture of the different stages of sleep.

work and social schedules, but it is there nevertheless. A person who is aware of this fact, and who learns to pay attention to his body rhythms, can know when he is most likely to function at his best or at his worst in the course of the day.

Today the teaching of the ancient Greek philosophers, "Man, know thyself," has taken on a new meaning. Thanks to the biological sciences there is much more we can know about ourselves than ever men could in the past. It is knowledge that can enhance health and happiness and reduce disease and distress, both in the life of the individual and in society as a whole. But whether and how widely knowledge of the tides of life will be applied remains to be seen.

Chapter 10
Mind Over Body

A mass of wrinkles, made up of some 10 billion cells and divided into two hemispheres each about the size of a fist—this is the human brain. How does it work, and to what extent has man learned to control it artificially? Electrical current is a very important factor, but how is it related to chemical action? Tests have shown how rats, once taught to stimulate the pleasure centers of their own brains, become so obsessed that they stimulate themselves to the point of exhaustion. Can this tell us anything about human beings? Man's ability is increasing with regard to altering mental states through electrode stimulation and drugs, and this demands full public awareness of the dangers of the "psychocivilized society."

Dr. José Delgado's professional colleagues protested that a bullring was no place for a scientific experiment, but the determined professor of physiology at Yale and Madrid Universities was not to be dissuaded from his plan. Besides, he considered it not so much an experiment as a demonstration. So into the bullring at Cordoba, Spain, he went, flourishing a matador's red cape. The huge bull saw him and charged. Delgado continued waving the cape, not even trying to lure the animal to one side. But when the bull was almost upon him the professor pressed a button on a small radio transmitter that he had in his hand—and the beast came to an abrupt stop in a cloud of dust. Delgado had known it would, for he had implanted receiver electrodes in the part of the bull's brain that controls motor activity.

The "brain sciences" have been one of the fastest growth areas in scientific research in the last two decades, especially in two main departments of research: the functions of control which Delgado demonstrated so dramatically, and the exercise of intellect and memory. To form a general picture of the human brain, imagine a pinkish-grayish mass of wrinkles, the consistency of set gelatin. It comprises some 10 billion cells, divided between two hemispheres each about the size of a fist. Of all the mysteries of living things, this organ, capable of storing the contents of a great library, is surely the most complex, most elusive, and most fascinating.

It is only in the last century, since the invention of the electron

Opposite: the ability to control animals through magic has been a preoccupation of human beings from the time of the Greeks. Here a man summons the birds of the world with a magic whistle, from a fairy story called "The Three Princesses in the Blue Mountain." Scientists of this century have refined their techniques to the point where they can control the movements of animals with even greater precision than if they had used magic.

Right, below right, and opposite: Professor José M. R. Delgado stops a charging bull in mid-flight by controlling an electrode implanted in its brain. He comments, "Brave bulls are dangerous animals which will attack any intruder into the arena. The animal in full charge can be abruptly stopped by radio stimulation of the brain. After several stimulations there is a lasting inhibition of aggressive behavior."

microscope and discovery of electricity, that intensive brain research has been possible. The greatest thinker of classical antiquity, Aristotle, believed that the function of the brain was to cool the blood. The leading philosopher of the scientific revolution in the 17th century, Descartes, taught that it was a machine operated on hydraulic principles, with pumps, pistons, and canals keeping vital fluids on the move. But being a religious man, Descartes could not propose a thoroughly mechanical physiology, so he suggested instead that the pineal gland in the midbrain area was the seat of the soul. There the brain sciences rested until the sophisticated use of microscopy and electricity revealed marvels of which the great philosophers had never dreamed.

It was known in the early part of the 20th century that the brain contained millions of active cells, the neurons, and the analogy commonly used to describe it was a telephone exchange. The brain was visualized as a network of electrical activity, with signals continually coming in and going out. Others saw it, as the British physiologist Sir Charles Sherrington picturesquely termed it, as an "enchanted loom." Another popular analogy was the executive section of a big company, with the managing director's office in the central cerebrum, the superintendents of motor activity (movement of legs, arms, tongue, and the like) in the cerebral cortex, the manager of the reflex actions in the cerebellum, the camera room located in the frontal lobes, and so on. Then came the age of the computer, and with it an even more comprehensive analogy for the brain and its functions was seemingly found. One analogy has succeeded another as man has tried to understand the mysteries of the brain, and all in their way are suitable, but none of them nor any combination of them can fully explain those mysteries.

The 19th-century pseudoscience of *phrenology* supposedly was a way to interpret the bumps and contours of the human skull for purposes of character analysis. It was crudely based on a principle that has, on the whole, stood the test of time—that distinct functions are localized in particular areas of the brain.

Delgado in the Bull Ring

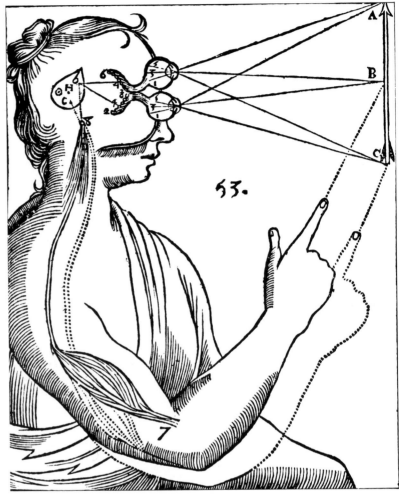

Above: René Descartes (1596–1650), French mathematician and philosopher. He suggested that the brain operated on hydraulic principles.

Above right: diagram from Descartes' book *De homine* (1662), demonstrating his mechanistic theory of brain function. Light from the object (A, B, C) enters the eyes and forms visual images (1, 3, 5) on the retina, which is connected to the walls of the ventricle by hollow tubes representing the optic nerve. The circular open ends of the tubes can be seen, and 2, 4, and 6 show the incoming or sensory stimulus. From the tubes the message goes through the ventricles by way of the animal spirits and reaches the pear-shaped pineal gland (H) which initiates the motor stimulus. Thus animal spirits from the ventricles are sent by way of the opening (8) into the nerve to the arm muscle which it inflates, producing motion. This is the basis of the reflex.

The "company headquarters" analogy incorporated this principle, which is graphically represented in Penfield and Rasmussen's drawing of the "Motor Homunculus of Man" reproduced on page 168. Other parts of the brain have been found to be associated with other functions, for instance the hypothalamus with the appetite, sex, pleasure, and pain; the hippocampus with memory; and the amygdala with fear and aggression. It is the increasing knowledge of these localizations that has enabled scientists like José Delgado to experiment with ways of controlling brain functions.

Delgado's electrodes implanted in the bull's brain simulated the normal electrical activity of the brain, which would have involved an impulse being flashed to the particular brain area concerned with inhibiting motor activity in the legs. Early in this century the telephone exchange analogy was considered an adequate explanation for this process, but throughout the 1930s and 1940s a prolonged debate took place as to whether the process of impulses jumping across the *synapses*—the gaps between brain cells—could be explained adequately in electrical terms. As it turned out, it couldn't; not only electrical but also chemical activity was involved in synaptic transmission. An electrical signal triggers a chemical mechanism which then squirts a "transmitter substance" across the gap. Since 1902 "chemical messengers" have been known to exist in the bloodstream and nervous system. Their discoverer, the British physiologist Sir William Bayliss, named these substances "hormones" in 1905, but until the 1930s few people realized that similar chemical transmissions took place in the brain. One of the few who did was the British neurophysiologist E. D. Adrian, whose valuable research on the subject was royally acknowledged when he was made Lord Adrian. One development from this research has been control of brain functions by the administration of chemical substances in appropriate doses—that is, by means of drugs.

In 1924 a Swiss psychiatrist, Hans Berger, found that if he pasted pieces of aluminum foil on his son's scalp and connected them by wires to a galvanometer, which in turn was connected to a pen to trace its fluctuations on a roll of paper, he could record

The "Company Headquarters"

Left: Sir Charles Sherrington (1857–1952), British physiologist who won the Nobel Prize in 1932 for his discoveries concerning the functions of the neurons. He introduced the method of analyzing reflex actions in experimental animals by the graphs of muscle contractions. He demonstrated the effect of reflex actions in enabling the nervous system to function as a unit.

"Alpha" Waves

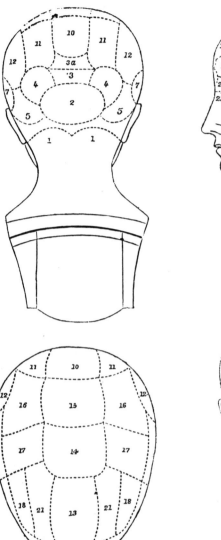

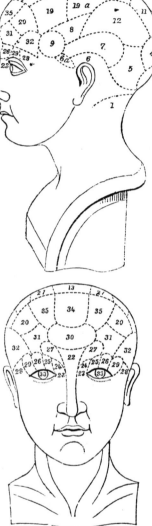

Right: diagram from *A System of Phrenology*, published in 1843. Different mental faculties were supposed to be situated in specific parts of the brain—for instance, "concentrativeness" (3); acquisitiveness (8); self-esteem (10); cautiousness (12); wonder (18); and individuality (22).

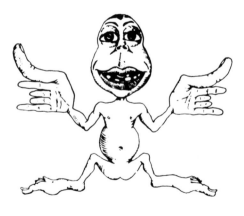

Above: Wilder Penfield's somatic-sensory homunculus, redrawn as a complete body showing the overrepresentation of certain particularly sensitive parts of the skin surface.

"brain waves," which altered according to the boy's state of mind. There was one type of wave which Berger called the "alpha," which was recorded when the boy was relaxed, in a wakeful but thought-free state. A slower one showed when he was asleep, and a faster one appeared when he was concentrating, for instance on a problem in arithmetic. Berger's discovery was virtually ignored at first, but when his findings were confirmed by Lord Adrian years later they were acclaimed as a breakthrough. There was speculation that one day it might be possible to read thoughts by deciphering the unceasing electrical chatter of the brain. Such expectations turned out to be unfounded, but the strip-chart record of brain activity, the *electroencephalogram* (EEG), proved a useful tool for monitoring man's most mysterious and most inaccessible organ.

Returning to the question of control, there are two main aspects: first, the degree of control exercised by the brain over the body and its functions, and second, the degree of control that one brain can exercise over another.

Left: Hans Berger, the Swiss psychiatrist who recorded the first brain waves in 1924.

Left: a page from Berger's notebooks, in which some of his classic experiments in electroencephalography are noted. He was a rather withdrawn man, and he did not publish any of his results until four years after his great breakthrough.

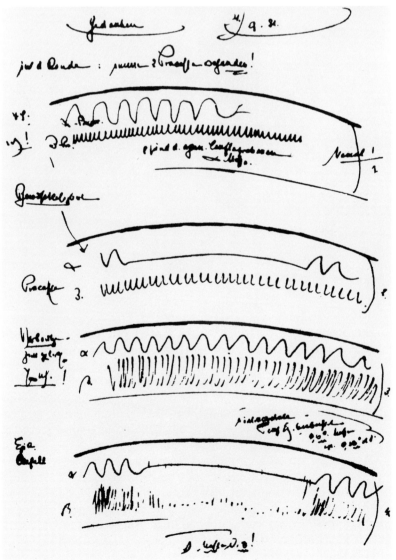

The Hypothalamus

The Italian physiologist Luigi Galvani got his name into dictionaries by observing in the late 18th century that frogs' legs twitched when they were electrically stimulated. A century later two Germans, Gustav Fritsch and Julius Eduard Hitzig, discovered that similar twitches could be induced in animals by applying electric current to the outer layer of the brain, the cerebral cortex, and that stimulation of different parts of the cortex caused different parts of the body to move. But not until 1924 was it discovered that emotions as well as motor activity could be influenced by electrical stimulation.

The discovery was made in Switzerland by Walter Hess, a Nobel laureate physiologist who was the first man to explore the functions of the hypothalamus. Experimenting with cats, Hess found that by stimulating one part of the hypothalamus he could

Right: engraved portrait of Luigi Galvani (1737–98), Italian physiologist born at Bologna. He believed that the source of the electric current he had discovered in frogs' legs was in the material of the muscle and nerve. In 1791 he formulated his theory of animal electricity.

Below: Galvani's experiments with frogs and electricity. He found that the muscles could be made to twitch by touching them, or the nerves connected to them, with a bridge composed of two rods of different metals. He concluded that inherent electricity was present in the frog. Another scientist, Allesandro Volta, realized that it was the junction between the two dissimilar metals which produced the electricity.

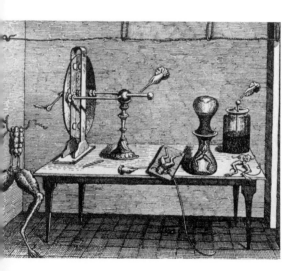

make them spit, snort, or growl, in short to show all the signs of readiness to fight, whereas if he moved his electrodes to a spot a fraction of an inch away, the cats exhibited quite different feelings, such as sexual arousal, hunger, thirst, fear, or drowsiness. Hess developed ways to implant electrodes permanently at specific points in the hypothalamus with great accuracy, and by attaching wires to the terminals on the cats' heads he was able to control their manifestations of emotion at will.

Hess' experiments provoked discussion as to whether the animals really felt the emotions they showed or were merely being forced involuntarily to fake them. But implanting and monitoring techniques were gradually improved and other experimenters did more work in the field. Eventually the question of the authenticity of the emotions became less important when the specific pain and pleasure centers in the hypothalamus were discovered and stimulation of them was actually observed.

In the 1950s a spate of experiments at several universities established beyond doubt the intensity of the feelings aroused by electrical stimulation of parts of the brain. The findings gave rise to alarmed speculation on the threat of this experimentation to the integrity of human personality. Researchers at Yale University in Connecticut implanted electrodes in cats' brains, then starved the cats until they were ravenous. When the animals were eventually given food they ate voraciously until the electrode in the pain center was activated, whereupon they dropped the food abruptly. Repetition of this treatment taught them to associate pain with the food, which they would then avoid in spite of being as hungry as ever.

The Yale psychologist B. F. Skinner, advocate of "behavior

Below: cutaway section of the human head and brain showing the hypothalamus (A) and pituitary gland (B). Working together, these glands control the hormone levels in the body and thus indirectly regulate all the body processes governed by hormones.

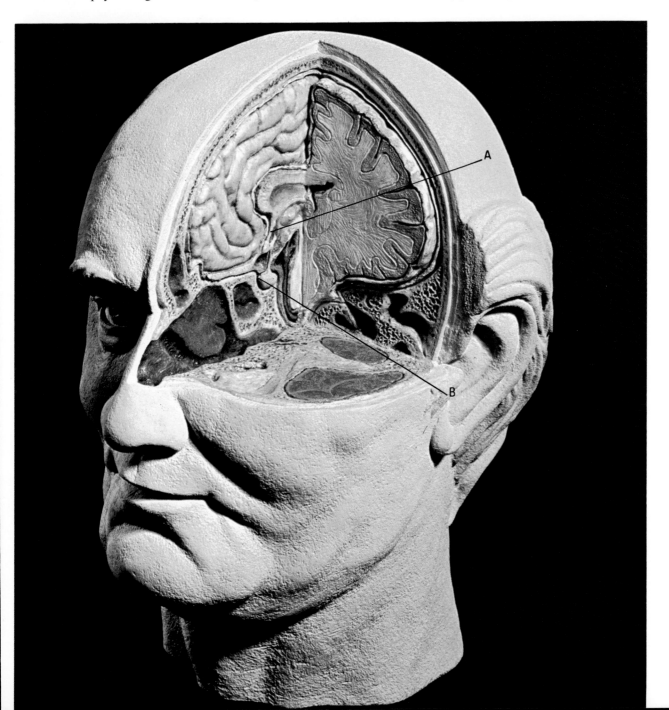

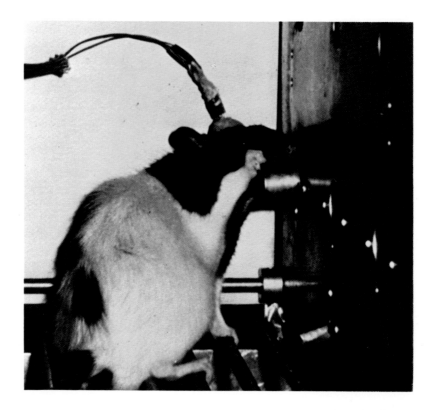

Right: a rat working a self-stimulating machine. During experiments it was found that, once rats learned to control the pleasure centers in their brains, they continued to stimulate themselves until they became completely exhausted.

modification" of human beings through the systematic administration of reward and punishment, devised a contraption that became known as the "Skinner box." It was designed so that the occupant of the box received a reward every time it performed a particular action, such as pushing a pedal. Generally the reward was in the form of food, but in 1954 James Olds and Peter Milner at McGill University, Montreal, discovered by chance the existence in rats of a hypothalamic pleasure center. One rat, apparently perversely, kept moving towards a particular corner of a maze instead of avoiding that corner as expected when its brain was electrically stimulated. Olds found the reason for this behavior in the fact that the electrode had been placed near the hypothalamus by mistake instead of in the midbrain area. He and Milner then carried out a controlled experiment, and the astonishing behavior of their rats was discussed throughout, and indeed beyond, the scientific world. When the rats learned to stimulate their own pleasure centers by pressing the pedal, they did so continuously. Caring nothing for food, sex, or rest, they would go on stimulating themselves for hours on end until they collapsed from sheer exhaustion. When they were taken away from the pedal, their longing to get back to it would spur them on to learn their way through mazes in record time. They would resolutely suffer shocks from crossing an electrified grid, which even starving rats would not cross to get food. Self-induced pleasure became their consuming obsession, and they would virtually stimulate themselves to ecstatic death.

José Delgado, the scientist of the bullring, played a prominent part in this type of research. His favorite subjects were monkeys and chimpanzees, for he was interested in experimenting with behavior modification in the context of a social group.

He worked with a monkey colony dominated by an aggressive and ill-tempered male named Ali. Ali would prowl around the cage while all the other monkeys cowered in a corner in terror of him. Then Delgado switched on the current to the electrode in Ali's brain, located so as to inhibit his aggression. The other monkeys soon learned to recognize when Ali's aggression was turned off, at which point they would move freely around the cage and even jostle him. Delgado then installed on the wall of the cage a lever that would switch on the current, and it was not long before a female monkey learned to manage Ali by rushing to the lever and pressing it whenever he threatened her.

A recent refinement introduced into this type of research is the use of a computer to modify and monitor a subject's mental states. Paddy, an experimental chimpanzee of Delgado's, had a box permanently fixed to his skull which the professor called a "stimoceiver," designed for the dual purpose of stimulating parts of Paddy's brain and receiving from it information about its activity. The box transmitted the information about Paddy's brain states, as reflected by his EEGs, to a distant computer programed to reward or punish particular mental states.

In one experiment with Paddy, the computer was programed to check for a spiky pattern in the recording of the chimp's brain-waves, and to administer a quick burst of electric current to his

Pleasure Centers

Below: monkey wearing Professor Delgado's stimoceiver. With the use of these small units, continuous radio stimulation and telemetry of the brain are possible while the subjects remain free to move and act as they wish.

Right: aggressive behavior is induced in the subject monkey by radio stimulation of the central gray area of its brain.

pain center whenever the pattern occurred. The spikiness was produced by activity in Paddy's limbic system (sometimes known as the "emotional brain"), and it occurred whenever he was aroused by anything, even a casual interest such as a new smell. At first the pattern occurred several times a minute, but when the computer remorselessly punished every manifestation of arousal, Paddy quickly learned to control his feelings. By the end of the experiment he was utterly subdued; he just sat around, automatically did whatever was required of him, and did not even dare take pleasure in his food.

The technology for controlling feelings and behavior through brain implants and a computer programed to react to manifesta-

Right: monkey pacified by radio stimulation of the caudate nucleus.

tions of specific mental states is available today, and there is very good reason to be concerned about the possibility of its abuse. Delgado himself is frankly and unashamedly enthusiastic about what he calls the "psychocivilized society." In his *Physical Control of the Mind*, published in 1971, he writes: "The procedure's complexity acts as a safeguard against the possible improper use of electrical brain stimulation by untrained or unethical persons." The implication that training in neurophysiology automatically produces ethical persons is obviously debatable—indeed, many people feel that the treatment meted out to experimental animals in the interests of research calls into question the ethical scruples of the researchers. A further safeguard against the possibility of the technology being abused may be the fact that, to achieve a "psychocivilized society," large numbers of people would have to submit to being implanted, and large numbers of doctors willing and able to do the work would have to be recruited, although the benefits of such a program might prove attractive to some governments.

There is probably greater cause for concern in the possibility of invasion of the brain and control of its functions by chemical means. In 1971 circumstances were brought to public notice which created a scandal, the issues of which are still controversial. Doctors in several American cities, notably Omaha, Nebraska, were found to be cooperating with school authorities in administering behavior modification drugs to some 300,000 American schoolchildren who were considered "hyperactive." The drugs quieted the children down, and, in the words of one enthusiast, "improved classroom deportment and increased learning potential." Little concern was shown either for the ethical implications or the long-term effects of this behavior modification program, which involved violation of the brain functions of a minority who were considered unruly in order to make them fit in with the requirements of the authorities and the prevailing socioeducational system.

Such violation is only possible in a society that has accepted in principle the use of psychoactive chemicals. The number of prescriptions for tranquilizers, sedatives, and antidepressant amphetamines written by doctors in any Western country today is staggering and overwhelmingly indicates our acceptance. Self-medication may be claimed as an inalienable right of an individual in a democracy, whether he wishes to stupefy himself or induce a mystical experience, but if this establishes the use of mind-altering drugs as normal, it will become that much easier to justify behavior modification through their use. But in fact the use of psychoactive chemicals for such purposes can be contested even without bringing in ethical considerations. Their effects on the brain and nervous system are too widely spread, variable, and dependent on other factors such as environment for them to be reliable in large-scale behavior-modification projects. An interesting experiment, reported by British psychopharmacologist C. R. B. Joyce in 1968, was carried out using two groups of 10 people each, who were kept in two separate rooms. Nine members of one group were given a barbiturate (sedative) and the

Paddy's Pattern

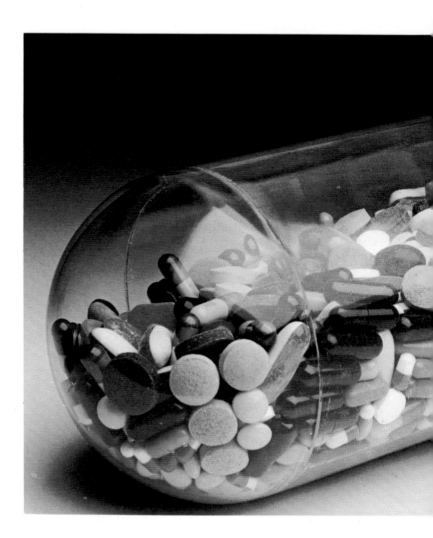

Right: the selection of tranquilizers, antihistamines, vitamins, aspirins, pep pills, stomach pills, headache pills, barbiturates, and so on that is easily available to modern man is an indication of how much we depend on pills and drugs to keep us healthy or cure our symptoms. Treating single aspects of a person instead of the individual as a whole is also a modern Western trend. The atmosphere thus created is one in which the disaster of thalidomide and horrors of drug addiction are logical "side effects."

tenth member an amphetamine (stimulant), while in the other group the distribution was reversed. In each case the tenth person, the odd one out, behaved exactly like his companions. The social pressure to conform with the group was apparently more "psychoactive" than the drug; indeed, the pressure not only countered but reversed the drug's effect.

More specific and predictable effects may be achieved with chemical substances which produce the ultimate in behavior modification—death. When the role of chemicals in synaptic transmission was discovered, biochemists got to work analyzing and attempting to isolate the chemical transmitters. In normal functioning of the brain, when a chemical transmitter has crossed the synaptic gap and done its job it is broken down by an enzyme. This process is essential to life, for otherwise the receiver cell would get clogged with an accumulation of transmitter substance and would cease to function. Obviously, effective weapons can be made from substances which, when introduced into the brain, are identical to the natural chemical transmitters in all respects except that they are invulnerable to destruction by the enzymes, or which poison or deactivate the enzymes themselves. South American Indians used to poison the tips of their arrows with a plant extract, curare, which brought about death through paralysis of the respiratory muscles

Psychoactive Drug Effects

caused by the blocking of brain synapses. Of course the Indians did not understand how the poison worked, but modern biochemists do, and many of them have been sponsored by governments to synthesize chemicals that would produce similar effects. This has not proved difficult, and today stockpiles of what are generally known as "nerve gases" are held by all major military powers.

Before we leave the modern horror stories, there is one other method of mind and behavior control that must be mentioned— that is, psychosurgery. In 1935 a Portuguese doctor named Egon Moniz developed a technique for "curing" violence by severing the nerves connecting the frontal lobes of the brain to its deeper emotional areas. The operations, known as lobotomies and leucotomies, became fashionable in the 1940s and 1950s, and were performed for the most part on nonconsenting patients in mental hospitals. A patient's husband, wife, or parent who gave consent often regretted doing so when back from the hospital came a person made not only nonviolent but also insensitive, mentally retarded, and lacking any capacity for either real enjoyment or forward planning. Extended studies of lobotomized and leucotomized patients resulted in the surprising discovery that the effects of the operations were not permanent. This in turn led to the conclusion that brain functions were not so

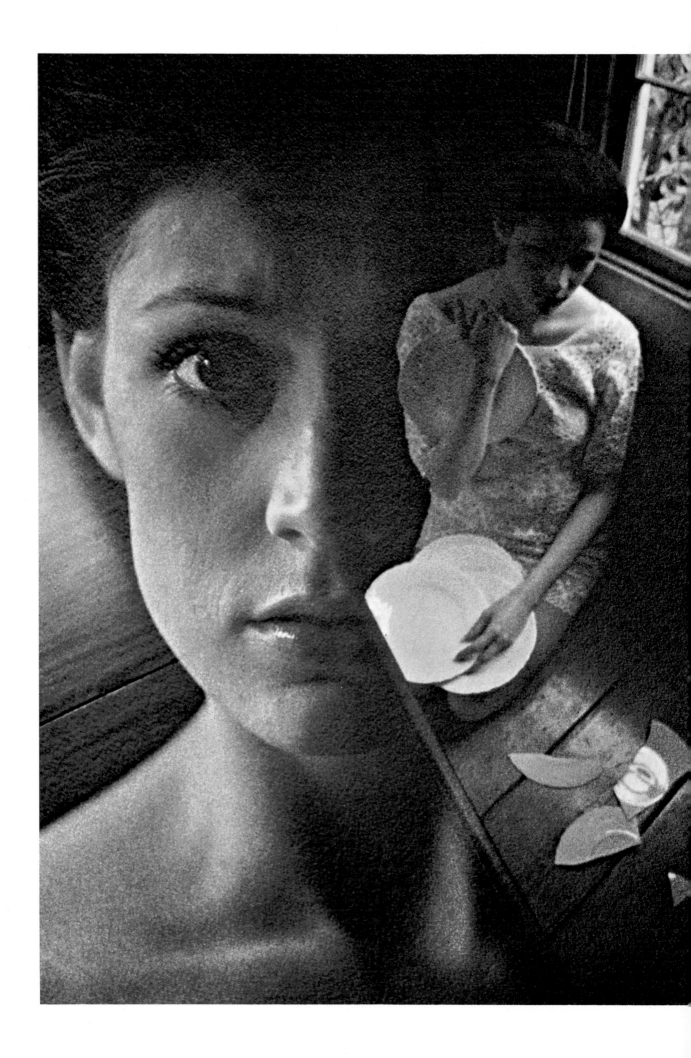

absolutely localized as had been thought, and that if a part of the brain were damaged or destroyed, its functions could eventually be taken over by another part. But despite its crudity and dubious effectiveness, psychosurgery still has its champions in the medical as well as law enforcement profession as a means of dealing with individuals who are criminally or insanely violent.

To consider another aspect of the question of control, it must be asked to what extent the mind can control physiological processes and events that under normal circumstances are involuntary. Breathing, heartbeat, temperature control, eye movements, digestion, blood flow, and many other processes are governed by what is known as the *autonomic nervous system*, a system of nerve pathways through the body made up of "visceral fibers." This term is used to distinguish them from the "somatic fibers," which serve the skeletal muscles and are under voluntary control. If man had deliberately to control these processes it would put intolerable demands upon his attention and greatly reduce his efficiency, so the question arises why he should ever try.

The idea that states of the body are directly influenced by states of the mind has gradually become a more and more familiar one throughout this century. It is well known that some physical illnesses are *psychosomatic*, and some people maintain that virtually all illnesses are. Although most medical care is still based on allopathy and surgery, an increasing number of modern doctors prefer to treat the body through the mind. Dr. Carl Simonton of Forth Worth, Texas, for instance, has reported the reversal of apparently terminal cancers when conventional treatments are combined with training in techniques of meditation and voluntary control of internal states.

In Germany in 1905 Dr. Johannes Schultz began to study different techniques of hypnotherapy combined with the disciplines of yoga. By 1910 he had developed a therapeutic system which he called "autogenic training." Through this system patients learned to relax tense muscles and regulate blood flow and heart rate by using verbal phrases, and eventually they could focus their efforts on controlling specific psychosomatic disorders. Schultz was a pioneer in this field, but although his method was highly successful in curing certain diseases it was never widely practiced. However, a modern development of it, known as *biofeedback*, is becoming increasingly popular today.

The basic idea behind biofeedback is simple. A person is made constantly aware of his involuntary or autonomic physiological processes so that he can learn to exercise control over them. In a typical biofeedback session a subject, sitting comfortably in a chair, had electrodes attached to the back of his head, his right forearm and two fingers of his right hand. He wears a special jacket equipped with a respiration gauge. An experimenter sits at a control panel behind him to direct the session. The aim is self-regulation of muscle tension, body temperature, and brain wave rhythm. The electrodes detect changes in these internal states and relay the information back to the subject by means of three bars of light on a screen in front of him. Each bar, like that of a mercury thermometer, becomes taller or shorter in response to changes in whichever physiological state it is monitoring.

"Biofeedback"

Opposite: states of the body are frequently influenced by states of the mind. Psychosomatic disease, fatigue, and mental illness can all be linked to the activity of the mind. Hypnosis, yoga, and biofeedback are all therapeutic systems based on this connection between the mind and body.

"Western Sufi"

Above right, right, and opposite: Jack Schwarz being tested in the Menninger research laboratories. Dr. Elmer Green and Alyce Green, who have taught the biofeedback system since 1964, are particularly interested in such subjects, who can show highly developed abilities of self-control. However, the main purpose of the center is to develop methods of self-treatment for ordinary people afflicted with asthma, epilepsy, or other potentially crippling diseases.

The subject first tries to achieve complete relaxation, during which he can watch his progress on the feedback meter wired to the muscle in his forearm. Having achieved this, he concentrates on temperature regulation. The electrodes attached to his fingers measure his success in raising his body temperature and relay the information through the second bar on the screen. He then tries to induce a state of consciousness that produces alpha brain waves, a state combining alertness with complete calm. The third bar rises to its maximum height when he succeeds in producing alpha waves for a period of 10 seconds.

In their psychophysiological laboratory at the Menninger Foundation in Topeka, Kansas, Elmer and Alyce Green have been experimenting with biofeedback techniques since 1964. They have seen some remarkable experimental subjects. An Indian yogi named Swami Rama demonstrated for them his control of the arteries in his wrist by simultaneously warming

up one spot on the palm of his hand and cooling down another spot only 2 inches away until there was a difference of 11°F between them. He then slowed down his heartbeat from 70 beats per minute to 52 beats per minute, taking less than a minute to make the change. Not content with this, he offered for the sake of the experiment to stop his heart from beating completely for three or four minutes, but Dr. Green said that an arrest of 10 seconds would be enough to prove his point. In fact, Swami Rama stopped his heart from pumping blood for 17 seconds by making it beat extremely rapidly, a state known as "atrial flutter," which normally results in fainting or death. Another subject, a "Western Sufi" from Oregon named Jack Schwarz, showed no physiological pain responses when burning cigarettes were held against his forearm for as long as 25 seconds. He could also drive a knitting needle through his biceps and control the bleeding when it was removed. Two seconds after he had said, "Now it stops," the bleeding ceased, and soon afterwards no sign of a wound could be found.

Men such as Swami Rama and Jack Schwarz have demonstrated the degree of voluntary control that man can train himself to exercise over his internal processes. But it is not the aim of biofeedback training to develop such sensational powers. Its primary purpose is therapy. The Greens report that their training has helped many different types of patient. Migraine sufferers have learned to alleviate their condition by practicing temperature control, which involves control of the vascular system and the flow of blood through the brain. People suffering from tension headaches have gained relief through muscle tension feedback. Swami Rama deliberately induced his atrial flutter, but people prone to the condition have learned to control it with the help of a biofeedback meter connected to an electrocardiograph. Epileptics, too, have benefited from biofeedback training, by learning to produce alpha brain rhythms. Even alcoholics and drug addicts have been helped, although no physiological information relevant to their condition can be fed

Right: Alyce Green and Dr. Dale Walters wire up Swami Rama for testing in the Menninger laboratories. Swami Rama was able to control the circulation of blood through his arteries and to slow down his heartbeat. Such abilities show what potential power the mind has over even the involuntary functions of the body.

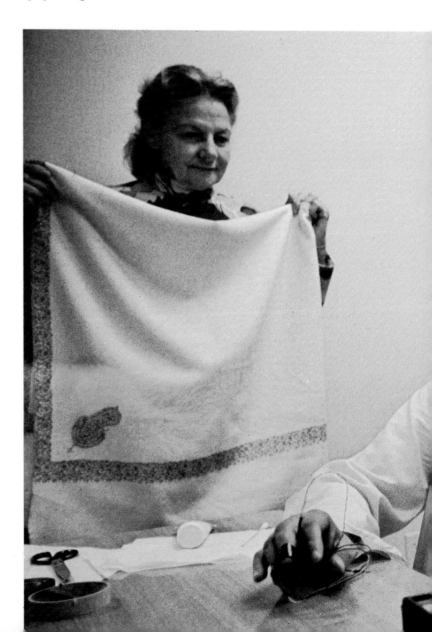

back to them through biofeedback instruments. In such cases it sometimes seems sufficient for the addict to learn through biofeedback training that he is not necessarily a slave to uncontrollable compulsions, but can in fact learn to master some of his inner processes. The resulting change in the subject's self-image has a therapeutic effect in itself.

Developments in physical science in the first half of this century have endowed man with awesome control over the forces of nature. Developments in the brain sciences in the second half are giving him equally unforeseen and far-reaching power to control his social institutions, his fellow men, and himself. While the possible consequences of the revolution in physical science range from nuclear holocaust to cheap energy for all, the consequences of the brain revolution range from a society of behavior-modified automata as portrayed by Aldous Huxley in his novel *Brave New World* to a society of free individuals enjoying the health and wisdom that come of self-knowledge and self-control. Science has a habit of confronting man with these situations, which are simultaneously a threat and an opportunity.

Swami Rama

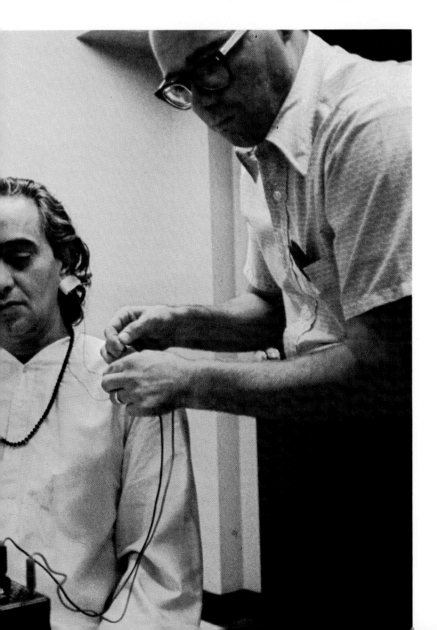

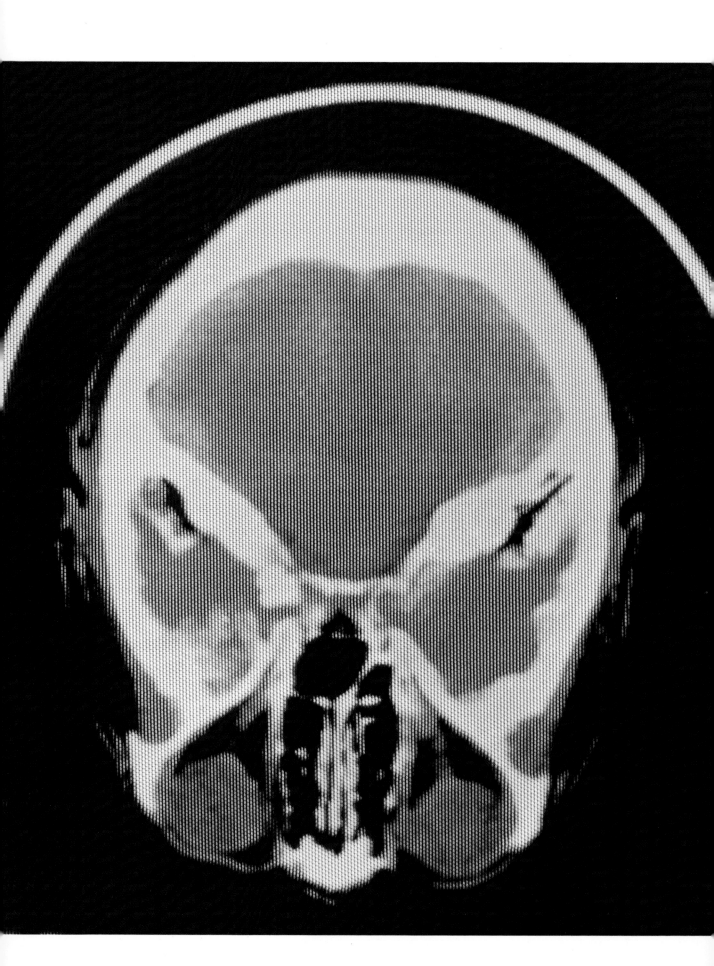

Chapter 11
Limits of the Brain

Is there any limit to the potential power of the human brain? In the search for a cure for epilepsy, researchers have made some remarkable discoveries of how memories are stored. Can learned skills be acquired by cannibalism—an untrained animal eating the brains of another which has been trained? What role does RNA play in memory formation? Human brains are "two-fisted," made up of two hemispheres, each of which controls different types of activities. What happens if part of the brain is damaged or surgically removed? Are computers really "superbrains" and humans just inferior computers, as some enthusiasts claim? The mysterious secrets of the human brain have yet to be fully exposed even today.

Rats, cats, and monkeys are probably as averse to people poking around in their brains as human beings are, but they can do less about it. That is one reason most of our knowledge of the living brain is based on experiments performed on animals. But animals don't talk, so it was not until researchers had the opportunity to probe the living human brain with electrodes that they learned of the prodigious capacity of the memory.

In the 1930s a Montreal neurosurgeon, Dr. Wilder Penfield, made some remarkable discoveries while working on the brains of epileptics. As the brain itself does not feel pain, it is possible to operate on it while a subject is fully conscious and therefore able to report what he experiences. Penfield was working on the theory that epileptic seizures are triggered by activation of a specific point in the brain, and his treatment consisted in locating the epileptic focus for the individual patient and destroying it. To locate the focus required extensive probing of the temporal lobe of the cortex with electrodes, and Penfield's surprising discovery was that when particular points were electrically stimulated people had vivid instantaneous recall of experiences long forgotten.

One woman, for instance, heard an orchestra playing an old popular tune whenever Penfield's electrodes touched a particular spot. It was as clear as if a radio had been turned on, she said, and she happily sang along with the music she was hearing until the electrode was removed (which she couldn't see or hear hap-

Opposite: a brain scan, one of the newest (and most painless) ways to explore the brain. The technique (computer-assisted tomography) is the only satisfactory method of examining soft tissue structures, and is capable of differentiating clearly between the gray and white matter of the brain. An X-ray tube and highly sensitive crystal detectors take almost 30,000 readings of tissue density which are transmitted to a computer. From this input an accurate, highly detailed picture of a cross section of the body—a "slice"—is projected on a television screen. Geoffrey Hounsfield, inventor of the machine, worked from the basic principle that different types of tissue in the body absorb different amounts of X-rays. The detector crystals replace the conventional X-ray plate. The prototype machines were first used on patients in 1970.

Brain Power

Below: Dr. Wilder Penfield (1891–1976), the American neurosurgeon who made important contributions to the knowledge of memory, speech, and motor and sensory functions. He performed experiments on epileptics in the 1930s which seemed to suggest that specific memories were stored at specific points in the brain which could be individually stimulated over and over using electrical impulses.

pening), whereupon she stopped abruptly. Another patient, a young man, was transported back to the boy's room at the school he had attended years before, and many others recalled in vivid detail sights, sounds, situations, and even long conversations from their past which they had supposedly forgotten. Penfield came to the conclusion that there is a mechanism in the brain which records everything that ever happens in a person's experience. When a touch of an electrode brought forth a memory, he wrote, it was as if "a strip of cinematographic film with soundtrack had been set in motion within the brain." The incidents were often quite trivial, the sort that seem to pass into oblivion soon after they happen, but the memory of each incident was related in such detail that it was for all the world as if the person were living through the experience at that moment. It seemed that

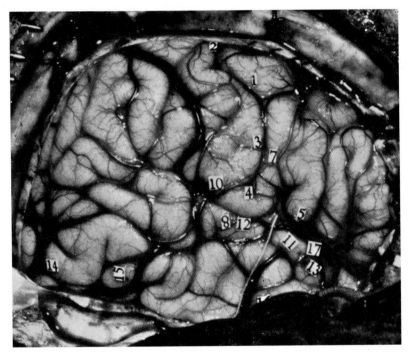

Above right: the right cerebral cortex of one of Wilder Penfield's patients. The points where stimulation produced positive responses are marked by numbers. The patient, a 26-year-old woman, was afflicted with recurring cerebral seizures.

there was really no such thing as forgetting—there was only failure to recall. The human brain, like a zealous clerk, filed away absolutely everything that happened. And every individual memory, it seemed, was lodged in a specific cell at a particular point in the cortex. Electrical stimulation caused the cell to "fire," and thus a memory was recalled.

But the solution to the mystery of memory was not that simple. As mentioned previously, people who have parts of their frontal lobe removed or disconnected through lobotomy or leucotomy operations have eventually recovered the functions or characteristics associated with the affected area. Observations of war casualties, and also the experiments of American neuropsychologist Karl Lashley at Harvard University, Massachusetts in the 1940s, suggested that memories were not so localized as Penfield's observations implied.

Karl Lashley patiently taught rats a variety of skills, such as finding their way through mazes, then operated on their brains and removed a part of the cortex. When the rats had recovered from the operation, he tested them to see if they still retained their learned skill. His expectation was that one of his removals would completely eradicate the rat's memory of the skill, thus making it possible to specify exactly where the memory was stored. But this didn't happen. The rats showed signs of damage to their memory as well as to other functions as they lost more and more brain tissue, but they still managed eventually to perform the activities they had learned before their operations. Memories, it seemed, were not localized but distributed throughout the cortex as a whole.

Both Penfield's and Lashley's findings were indisputable, but they were also apparently irreconcilable. If memory was depen-

Above: Karl Lashley (1890–1958), American neuropsychologist noted for his pioneer investigations of the relationship between brain mass and learning ability. He performed experiments in which he measured learned behavior before and after causing specific, carefully quantified brain damage in white rats.

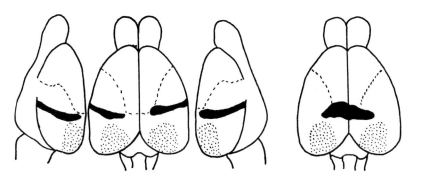

Above left: drawing of the brains of Lashley's experimental rats. Lesions (dark areas) were made so as to separate partially the visual area (dotted) from the motor areas (outlined by broken lines) of the rat's brain; they did not disturb visual learning.

dent on the electrical activity of the brain—on neuron networks and their connections—as had long been assumed and as Penfield seemed to have demonstrated, then it must therefore be localized. Lashley had assumed that it was, and his own research brought him to the despairing conclusion, as he wrote in his last scientific paper in 1950, that "learning just is not possible at all." However, in the 1960s discoveries were made which suggested that memory might not be an electrical but rather a biochemical phenomenon. The discovery of DNA and RNA molecules and their capacity for storing genetic information had led to the use of the term "genetic memory." To say that a gene "remembered" was simply a convenient metaphor for its informational storage capacity. Or was it? It would be ironic if the borrowing by geneticists of a metaphor from the brain sciences should in turn provide the clue to one of the brain scientists' own basic problems. This thought occurred to the Swedish chemist and researcher Dr. Holger Hydén in the late 1950s.

Hydén first trained rats to perform a balancing feat then found that when they had learned to do it the cells in the part of their brains that controlled balance had 20 percent more RNA than before. This extra RNA constituted the basis for the formation of new brain protein. Perhaps the memory of the newly-learned skill was associated with the new protein. His hypothesis was supported by the research of Professor David Krech and his colleagues at the University of California at Berkeley in 1959. The Berkeley team reared two groups of rats. They kept one group adequately fed but separated from each other, idle, and

The "Mau Mau" Hypothesis!

mentally bored, while the other group was brought up in an "enriched" environment full of stimulating playthings, challenges, and plenty of space. After spending some three months in their separate environments, the rats were killed and their brains dissected. The researchers were in fact looking for chemical differences between the brains of the two groups of rats, but they were astonished to find that the cortexes of the rats reared in the "enriched" environment were on average 4 percent heavier than those of the others. The discovery seemed to bear out the old and commonly discarded belief that intellectual activity made the brain grow.

Krech continued his experiments with two groups of rats selectively bred over generations to be "maze-bright" and "maze-dull" respectively, that is, either gifted or not in finding and memorizing their way through a maze. He found that when the gifted rats were kept in an impoverished environment and the dull ones in an enriched environment their maze-running skills became about equal. When the gifted were given the added advantage of being reared in the enriched environment, the difference between the two groups doubled. Applying these results with rats to human beings, Krech summarized the implications of his findings as follows: "We can now undo the effects of generations of breeding. Heredity is not enough. All the advantages of inheriting a good brain can be lost if you don't have the right psychological environment in which to develop it."

Making this type of logical leap from rats to human beings may be justified by the fact that both species have a cortex, or learning area, in the brain. But extrapolating from worms is rather more dubious, and fortunately so, for otherwise Dr. James McConnell's experiments might be taken as an incentive to cannibalism.

McConnell, a psychologist at the University of Michigan in the 1960s, decided to use planarians, tiny fresh-water flatworms, as his experimental subjects because they are the simplest creatures having rudimentary brains and synapses. His initial working hypothesis was that learning was a synaptic function. By administering a slight electric shock to his worms whenever the light was switched on, he trained them to react to light. Since worms are remarkable for their ability to grow new heads when cut in half or into pieces, McConnell wondered whether these newly-generated heads would remember lessons learned by the original whole worm. He found that they did, and he came to the conclusion that memories must be stored in cells distributed throughout the worm's body.

Testing this idea brought McConnell to his most sensational and most controversial conclusion. Knowing that planarians can turn cannibalistic, he starved a number of untrained flatworms for weeks and then fed them on chopped-up pieces of other planarians, some of which had been trained to react to the light while the rest had not. "To our delight," McConnell reported, "the cannibals that had eaten educated victims did significantly better . . . than did cannibals that had eaten untrained victims. We had achieved the first interanimal transfer of information, or, as I like to put it, we had confirmed the Mau Mau hypothesis!"

Other scientists repeated McConnell's experiments and those of one of his students, Allan Jacobson, but at first without success. About a year after McConnell's announcement 23 of them reported their negative findings in a scientific publication. However, one of this group later came up with results which confirmed McConnell's. He was Dr. William Byrne, a biochemist at the University of Tennessee Brain Research Institute, and he had found that if young rats were injected with RNA taken from the brains of trained rats they showed evidence of "remembering" the donor's training. Byrne's finding made sense to McConnell, who, despite his "Mau Mau hypothesis" joke, had always maintained that memories could not be transferred by cannibalism among higher animals because their digestive systems were too efficient in breaking down food. But if RNA extract were transferred direct from brain to brain, it was conceivable that memories might be transferred in the higher animals, too. McConnell repeated Byrne's work successfully, but then in 1972 he suddenly gave up work on memory transfer, claiming that it would take another decade before the scientific establishment acknowledged the truth of the principle.

Research into the "memory molecule" hypothesis has been continued by Dr. Georges Ungar at the University of Texas in Houston. Ungar taught mice to fear the dark, which is contrary to their nature. He then injected brain extract from trained mice into others who were untrained and found that the fear was transferred. He then isolated a chemical component of the brain extract which he called "scotophobin" (Greek for "fear of the dark") and asked a chemist colleague to synthesize it. Ungar then injected the synthetic scotophobin into other untrained mice and reported that they showed the same fear reaction as both the original trained mice and those given natural scotophobin.

Ungar's research has been dismissed by many orthodox scientists as a hoax or a fantasy, while enthusiasts, on the other

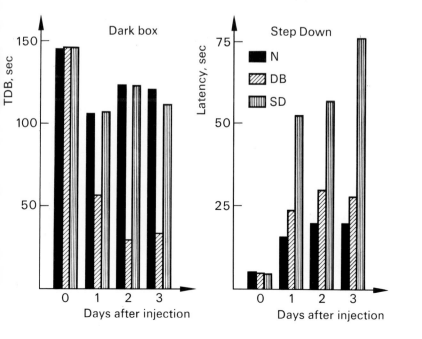

Left: results of Ungar's experiments in the cross-transfer of passive avoidance. Two tasks studied were dark-box avoidance and the use of a step-down platform. Mice were injected with extract from the brains of other mice who had been trained either to fear and avoid a dark box or to step down from a platform. Half the mice injected with dark-box extract (DB) were tested in the dark box, and the other half in the step-down situation. The group of mice injected with step-down extract (SD) were similarly split for testing. The results show that significant positive transfer was obtained when the mice were tested in the situation corresponding to their injections, but little or no transfer was noted when they were tested in the other situation. The numbers at the side of the graphs denote time spent in the dark box (in seconds) and time spent refusing to use the step-down platform.

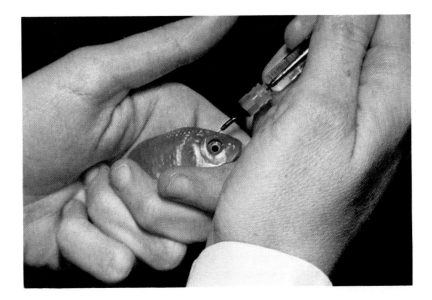

Below and right: Dr. Bernard Agranoff's experiments with goldfish and puromycin. After training goldfish to perform a specific action, Agranoff immediately administered a brain injection of puromycin, an antibiotic which halts protein formation, to half of the trained fish. Subsequently the injected fish seemed to retain no memory of their acquired skill, thus suggesting that memory may be linked to protein formation in the brain.

hand, have speculated that eventually specific memory molecules will be manufactured to order in the laboratory and either used to enhance learning or as a diabolical method of behavior or personality modification. However, since even Dr. Ungar's highly controversial claims do not go beyond the possibility of synthesizing a memory molecule of a conditioned reflex, it is certainly fantasy at this stage to imagine that the infinitely more complex memories involved in human learning might be transferred or manufactured.

But it may be possible to improve memory functions biochemically. Ewen Cameron, a psychiatrist from McGill University in Montreal, tried giving elderly hospital patients with memory difficulties massive doses of a yeast RNA preparation, and he found that their performance in memory tests improved significantly. Other researchers have experimented with inhibiting memory formation, hoping thereby to learn more about how new memories become fixed. Dr. Bernard Agranoff, a biochemist at the University of Michigan's Mental Health Research Institute, experimented with goldfish, training them to swim to one end of a tank whenever a light was flashed. When the training, done by means of mild electric shocks which followed the flash, was completed, Agranoff immediately gave half the goldfish a brain injection of puromycin, an antibiotic that stops protein formation. When the fish were returned to the tank, the injected ones did not swim away when the light was flashed as did the untouched ones. They had forgotten their lessons, and their lack of memory was apparently linked to the inhibition of protein formation in the brain.

The obvious question raised by this research was whether chemical stimulants, applied in the same way, might have the opposite effect of facilitating memory consolidation. In the late 1950s Dr. James McGaugh, a psychobiologist at the University of California at Irvine, put this question to the test of controlled experiment. Using various stimulants, such as strychnine and amphetamine, he found that learning was accomplished by experimental animals with greater speed and efficiency if they were given an injection soon after the learning experience. The sooner the better, in fact—the effect diminished in relation to the time element, and if too long a period elapsed then no effect attributable to the drug was noticed. So memory formation seemed to be a time-dependent phenomenon connected with protein synthesis in the brain, and overall the research suggested that selective obliteration, enhancement, or retardation of memory formation might eventually be accomplished with drugs. Such a technology would have obvious applications of interest to government agencies, such as in espionage work. Dr. Agranoff, when asked by a reporter whether the C.I.A. had ever approached him in connection with his work on memory inhibition, replied with a smile, "I forget."

A fact sometimes overlooked by zealous brain researchers is that human memory performance can be greatly improved without recourse to chemical agents. Our forebears must have had better memories than we do, as many of them could recite whole books, for instance the scriptures or epic poems. Before the age of printing, a trained memory was even more of a professional asset than it is today. Frances A. Yates, a Renaissance scholar, in 1967 published a study of ancient memory systems in her book *The Art of Memory*. It appears that the basic principle is to place whatever is to be remembered in a familiar scene, such as a street or geographical location, so that it can be recalled in sequence by association. Stage "memory men" use similar techniques. As noted earlier, Penfield's research led him to the conclusion that all a person's experiences are recorded and filed away, but most are placed beyond recall. Memory training—both as practiced by our forebears and stage performers and as taught by numerous correspondence colleges throughout the world today—is primarily training in techniques of recall. The problem is in getting information out of storage, and these systematic techniques are somewhat similar to the retrieval systems used in modern computers. This is one reason the computer is often used today as a model of the brain.

Memory feats, memory enhancement experiments, and comparisons with computer capabilities all suggest that human brains *could* function far better than they do. It is often said that we use only a fraction of our brain, with the implication that if we really tried we could bring more of it into operation, thereby becoming supermen or capable of solving crucial, seemingly insoluble world problems. This suggestion and its implications are widely believed and are therefore worthy of closer examination.

The idea that man uses only a fraction of his brain is based on the fact that the "association areas" in his cortex are proportionately much larger than in other species, except perhaps cetaceans (whales and dolphins). The association areas have no direct connection with sense impression input from the outside world. Their function seems to be to act upon information already received. The mental functions of discrimination, appreciation, creativity, and intellection take place in the association areas, and these areas are so large in man, while the differences between the mental powers of individuals are also so great, that it is natural to assume the existence of a tremendous amount of spare capacity or unused potential.

Testing Memory

Below: Greek bard or storyteller, also called a rhapsodist. These professional reciters of epic poems traveled from court to court and town to town, repeating by memory long poetic tales of heroes and gods. This figure comes from the neck of a Greek amphora or vase found at Vulci and dating from 490–80 B.C.

Above: *The Boyhood of Raleigh* by the 19th-century British artist Sir John Everett Millais. Storytelling has always been a proof of a fine memory, which is primarily acquired by practice in techniques of recall.

On the other hand, both ablation experiments with animals, in which parts of their brains are removed, and studies of brain damage in humans suggest that the brain's spare capacity is meant for maintaining rather than enhancing its efficiency. People who have had substantial portions of their brains surgically removed have often suffered only temporary loss of the functions associated with the removed areas. In fact, a capacity to house extra "spare parts" appears to be a characteristic of the brains of higher mammals. A Chicago biologist, Jack Cowan, has demonstrated that to insure reliable performance of a machine made up of many parts over a long period of time, any part of which might fail at any time, the best arrangement is for the parts to be connected randomly, allowing many alternative pathways and a high rate of overlapping functions among them. There are many facts which support this model of the brain's capacities and functions. There is the return of functions primarily associated with ablated or removed areas, as well as the fact that although millions of brain cells die off daily in each of us, we do not generally as we get older suffer random failures of brain functions. Indeed, the very old often have marvelous long-term memory, which we would not expect if their ever-vulnerable brain cells held specific and unique contents which could be lost for ever.

The "redundancy" theory of the purpose for the brain's spare capacity also makes it possible to reconcile Penfield's finding that memories are localized with Lashley's that they seem to be distributed throughout the cortex. A specific memory may be stored many times over in different parts of the cortex—the fact that Penfield's electrodes repeatedly applied to the same spot fired the neuronal activity that led to recall of a specific memory does not necessarily imply that the memory is stored only in that spot. In fact, considering the fact that brain cells die off in millions and at random, it is unlikely that the brain could be the highly efficient organ it is if it didn't carry duplicates of its contents and wasn't able to transfer the control of essential functions from one cortical area to another.

The brain's spare capacity, therefore, is probably not an indication that man has fallen far below his evolutionary potential, but rather a property of the brain itself that makes for optimum efficiency over a long period by allowing for accident and loss. On the other hand, it is nevertheless quite true that most people do not extend themselves in the use of their brains, and for various reasons—social, psychological, personal, physical—immense potentials go unfulfilled.

Jack Cowan's model of the human brain is, of course, based on comparison with computers, a concept often used today as a replacement for the old "company headquarters" and "telephone exchange" analogies. To what extent is this relevant? Are computers really "superbrains" and human beings inferior computers, as some enthusiasts believe?

At first sight there are quite remarkable parallels between brains and computers. Computers have information-input terminals that correspond with human senses. They have microscopic transistors with multiple interconnections like brain neurons, they have short- and long-term memory capabilities, and they have a main unit for processing and acting on information received, just as the brain's cortex does. It used to be said that if a computer of brainlike capability were to be built, it would have to be as big as London's Royal Albert Hall or the Empire State Building in New York City. But the technology of miniaturization has progressed so rapidly in the last decade or so that the latest estimate is that it could be fitted into a fair-sized room. But even if it were conceivable that such a machine could be shrunk to the size of two clenched fists, would it be a brain? Current discussion of this question tends to be passionate and heated, but a reasonable answer might be that it would far surpass the brain in its capacity for rapid calculation and for reliable and rapid retrieval of information stored in its memory. If these are the main and characteristic functions of brains, then this computer would indeed be a superbrain. But this last point is really the crux of the matter, and it is too often left out of the discussion.

As computer intelligence has advanced tremendously in the last two decades, and is still advancing, perhaps we should not be too complacent about the superiority of the intelligence functions of our human brains. On the other hand, the possibility that man may eventually become subordinate to a super-intelligent computer of his own making seems at present a baseless fear. Anyway, this would require a sociopolitical revolution as well as fantastic future progress in computer technology. Today computers can be programed to play an excellent game of chess, to translate technical literature from one language to another, even to conduct conversations with themselves on specific topics which give the impression of a dialogue between two independent minds. In the forseeable future computers may well be used as teachers, as medical diagnosticians, as land, sea, and air traffic controllers, and as economic, political, and military advisers and strategists, thus taking on—some would say usurping—many of the functions of human brains. But for the generation of creative ideas, both in the sciences and in the arts, and above all for the integration of such noncomputable factors as love, compassion, and morality with thought, human brains will surely remain supreme and unique. It is even doubtful whether computers will in the future be used to their full potential for other than routine tasks. As an anonymous naval officer is reported to have put it, "Man is destined to remain for a very long time the lightest, most reliable, most cheaply serviced and the most versatile general-purpose computing device made in large quantities by unskilled labor."

One particular characteristic of the human brain distinguishes it both from other mammalian brains and from computers—what is technically known as *asymmetry*. Only quite recently has the fact that the two distinct hemispheres of the brain have different functions and different modes of thought been discovered. The hemispheres are joined by a thick cable of white matter known as the *corpus callosum*, and occasionally it is necessary to cut through this for medical reasons. Such surgery has been found effective in treating chronic epileptics, since it results in the epileptic focus point being localized in one half of the brain so that the other half can bring the seizure under con-

Brains v. Computers

Below: an IBM portable computer designed to fit on an office desk. The technology of miniaturization has progressed so far tnat the multiplicity of circuits and space for storage of information can be accommodated in this desktop machine. More complex computers which can perform more tasks, understand more "languages," or store more information would probably need to be larger.

Below: visual input to the bisected brain was limited to one hemisphere by presenting information in only one field of vision. The right and left fields of view are projected, via the optic chiasm, to the left and right hemispheres of the brain respectively. If a person fixes his gaze on a point, therefore, information to the left of the point goes only to the right hemisphere and vice versa. Stimuli in the left visual field cannot be described by a split-brain patient because of the disconnection between the right hemisphere and the speech center, which is located in the left hemisphere.

Below right: visual-tactile association performed by a split-brain patient. A picture of a spoon is flashed to the right hemisphere; with the left hand he retrieves a spoon from behind the screen. The touch information from the left hand projects mainly to the right hemisphere, which means the patient is unable to say (using the left hemisphere speech center) what it is he has picked up.

trol. In 1961 a war veteran who had had this operation, who is known in the scientific literature as "W.J.," was put through a series of tests by Roger Sperry and Michael Gazzaniga at the California Institute of Technology. Some bizarre observations as well as some very interesting facts emerged.

It had long been known that the right brain hemisphere controlled the left side of the body and vice versa. In his first tests, W.J. was able to carry out verbal commands like "Raise your hand" only with the right side of his body, which suggested that only the left hemisphere understood the command. This implied that normally the two hemispheres receive information in different ways and transfer it between themselves through the corpus callosum in order to coordinate the actions of the two sides of the body. Further tests bore this out. If his right eye was covered and he was shown a familiar object such as a spoon, W.J. was unable to name it, but if he was given it to hold he could name it instantly. So the name must be stored in the left half of the brain, the half disconnected from visual input because of the blindfold. Speaking, reading, the performance of tasks requiring judgment or interpretation based on language, these were all done by the left hemisphere of the brain. The right hemisphere was mute and apparently imbecilic.

Or so it seemed. But further tests with W.J., and later with other split-brain subjects, showed that the right hemisphere had its own abilities and talents. W.J. could copy a drawing rapidly and accurately with his left hand but not with his right. Given some colored blocks to arrange according to a diagram, he could do so without difficulty with his left hand, but again his right hand proved incompetent, and when both hands tried to work at the task simultaneously the result was confusion, with the right hand disarranging the blocks as fast as the left hand

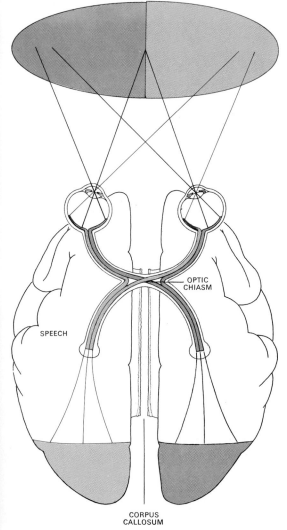

OPTIC
CHIASM

SPEECH

CORPUS
CALLOSUM

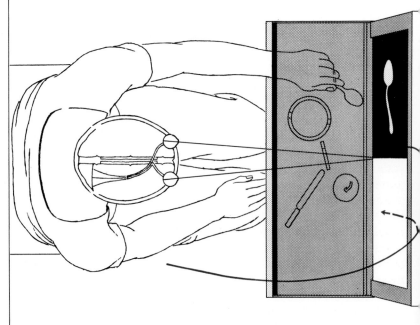

EXAMPLE	LEFT HAND	RIGHT HAND

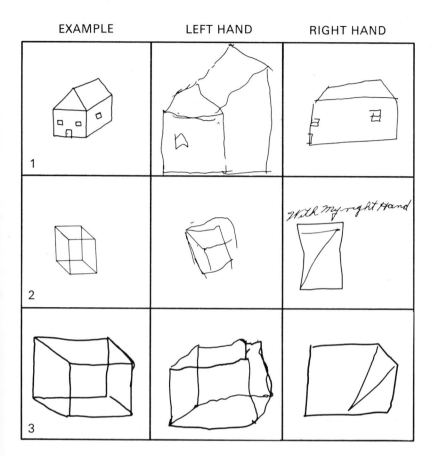

Hemispheres of the Brain

The human brain is actually two brains or hemispheres connected by an isthmus of nerve tissue called the corpus callosum, and when the cerebrum is divided surgically each hemisphere functions independently as if it were a complete brain. The American psychologist Michael Gazzaniga has investigated the split brain in depth.

Left: Gazzaniga found that "visual-constructional" tasks are handled better by the right hemisphere of the brain. This was seen most clearly in his first subject, "W.J.," who had poor control of his right hand. Although right-handed, he could copy the examples only with his left hand.

arranged them. Obviously, the right hemisphere was superior to the left in spatial abilities.

Earlier in this century a Russian neurologist, A. R. Luria, reported the case of a certain composer who had suffered a stroke which had affected his left hemisphere, thereby paralyzing the right side of his body. He was subsequently able to compose better than before, although he was no longer able to write musical notation. Here was a dramatic illustration of a point that split-brain research is continually confirming—that although the right half of the brain is illiterate and mute, it is apparently the source of most of what we consider the higher and distinctive attributes of human beings: intuition, wisdom, and artistic creation. One of the pioneers of computer technology, Joseph Weizenbaum of the Massachusetts Institute of Technology, has made the point in his wise and passionate book *Computer Power and Human Reason* (1976) that it is precisely such right-hemisphere functions that machine intelligence will never be able to simulate. Another might be wit. It is said that an earnest programer once fed into his computer the question, "Is there a God?" and the machine delivered the print-out answer, "There is now." Could people ever seriously believe this story? Surely not. As Weizenbaum writes, "Folk wisdom knows the distinction between computer thought and the kind of thought people ordinarily engage in." And although the human brain may not be such a fast "number-cruncher" as a computer, its functions are more varied, its mysteries deeper and more elusive, and its capacity for the creation of beauty, wisdom, and truth noncomputable.

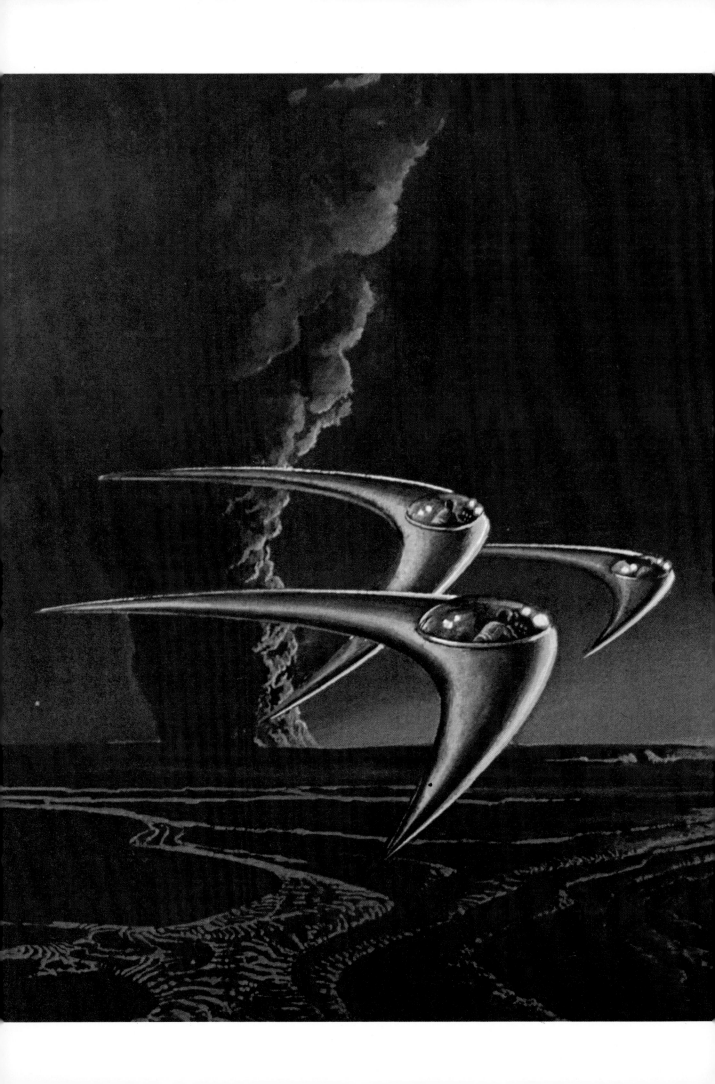

Chapter 12
The Question of
Life Elsewhere

The basic chemicals necessary to support life are abundant on other planets, both in our solar system and in others. Many scientists are convinced that life, and probably intelligent life, must exist in places other than on earth. One astronomer estimates that there may be over 100 thousand inhabited planets in our galaxy alone. What forms might extraterrestrial life take? Science fiction writers have put forward many ideas—could any of these be prophetic? What conditions would be required for the evolution of life as we know it—and how practical is the physical shape of man? Is life as we know it the only possibility?

In 1864 a large meteorite fell to earth near Orgeuil in the south of France. Chemists from around the world analyzed material from the meteorite, and it was announced that it included organic matter. Did this not prove, some scientists asked, that there must be organic life elsewhere in the universe? France's own great chemist, Louis Pasteur, also investigated. Well aware of the possibility of contamination by microorganisms from the earth's environment, Pasteur had a special drill made so he could take samples from the interior of the meteorite. He then subjected these to tests, designed to stimulate the growth of any microorganisms that might already have been present in the material, but his results were negative. It seemed that all the evidence of life which other scientists had found in the meteorite had probably come from earth.

With the development of more powerful analytical techniques, the study of meteorites has continued to the present day. Before the technology capable of sending manned or instrumental probes to other planets was perfected, no other possibility existed for testing extraterrestrial material for signs of life. As one scientist said before the first probe was dispatched to Mars, to discover even microorganisms on another planet would change the concept of "life" from a miracle to a statistic. Considering that there are billions of planets in the universe, the presence of the simplest organic life forms on our nearest neighbor would imply that there must be more highly evolved forms elsewhere. To

Opposite: an artist's conception of conditions on another planet. This painting, by Bruce Pennington, illustrates *The Big Sun of Mercury*, a novel written by the prolific science fiction writer Isaac Asimov. Ever since authors such as Jules Verne began to speculate on conditions in parts of the world they could not see, hosts of books, movies, comics, and illustrations have been produced which speculate on the appearance of extraterrestrial life.

Right: pieces of the Orgueil meteorite which fell to earth in 1864. They have been analyzed many times in search of evidence for life in space.

Below right: the Barwell meteorite. This large piece of rock fell to earth at Barwell, Leicestershire, England on Christmas Eve 1965.

Above: crater made by the impact of a piece of the Barwell meteorite.

discover simple life forms in meteorites would be just as suggestive. Some excitement was aroused in scientific circles a few years ago, when American microbiologists discovered something in the Orgeuil meteorite that seemed to be of biological origin but to resemble no known earthly organism. A few months after the discovery, however, other American scientists were able to show that the mysterious something had a structure identical to that of ordinary ragweed pollen. The meteorite has remained in a French museum for over a century while the question whether life might exist elsewhere in the universe remains unanswered, just as it has remained unanswered by the instrumental probes of the Martian environment of the 1960s.

Many scientists are nevertheless convinced that life, and probably what we would call intelligent life, must exist elsewhere than on our earth. Astrophysicists have established beyond doubt that our earthly laws of physics and chemistry hold good through-

Meteorites and Astronauts

Below: U.S. astronaut Edwin Aldrin of the July 1969 Apollo 11 moonshot. One of the most important aspects of the scientific work done by the two astronauts who walked on the moon was to collect rock samples. Aldrin told Houston Mission Control that the moon's surface was rather slippery—the fine dust particles of the powdery lunar surface were easily disturbed and clung to his spacesuit and boots.

Informed Guesses!

out the universe, and that the chemicals necessary for life are abundant on other planets of our own solar system as well as of other systems. Astronomers estimate that there are at least 10^{20} (that's 10 with 20 zeros after it) stars in the visible universe. The American astronomer Harlow Shapley has worked out some interesting figures and conjectures beginning with this premise. Suppose that only one in 100 stars is circled by planets following stable orbits; then suppose that only one in 100 of these has a planet like earth, and that, of all the earthlike planets, only one in 100 is situated so that it is neither too hot nor too cold for organic life to flourish; suppose finally that only one in 100 of these planets has a surface chemistry suitable for sustaining life. That would still leave 10 thousand million habitable planets in the universe. This is actually a rather modest calculation, for it would mean only one habitable planet per galaxy. The astronomer Frank D. Drake of Cornell University, New York, has estimated that there may be over 100 thousand inhabited planets in our galaxy alone. Other scientists have made similar "informed guesses" and come up with different estimates, but for all their

Right: an inhabitant of the moon, conceived and drawn by Georges Mélies, pioneer French film director. *"Les Sélénites"* are only one visualization of the life forms which many people hoped would be found on the moon.

differences they do agree on one thing—that life is probably a fairly common occurrence in the universe.

The question whether life elsewhere would be the same as life as we know it has been given a good deal of attention, too. Science fiction novelists have considered themselves free to speculate on the subject since there is no hard empirical evidence one way or the other. Many of these works have been mere uninformed fantasy, but some, such as the 1937 novel *Star Maker* by British author Olaf Stapledon, have been based on relevant scientific knowledge and evolution theory. Such books can stand as useful speculative contributions to *exobiology*, the science concerned with the biology of extraterrestrial life, which has been described as the only science that still must prove the existence of its subject matter.

Stapledon's space traveler visits many different worlds where life has evolved into a diversity of bizarre forms. On the first inhabited planet he visits, conditions are very similar to those on earth, and the inhabitants are basically of humanoid form but with differences. They have long necks, spindly legs, green hair,

Below: *Humming Ship*, another view of life forms to be found on other planets, painted by Tim White in 1976. Anatomical structures different from those on earth, including nonbiological forms, have been suggested by various writers, artists, and scientists.

spoutlike mouths, hands with three fingers, a thumb, and no palm, and eyes located below the nose. In these beings the senses of sight and hearing are underdeveloped compared with man's, but the senses of taste and smell are extremely acute. As they have taste receptors in their hands, feet, and genitals, they interact with each other and with their environment quite differently from human beings. But biologically they are very close to man, compared to some of the other creatures the space traveler encounters on other worlds.

On one world envisaged by Stapledon, the dominant intelligent organism has developed from a five-pronged marine animal like a starfish. It uses four prongs for movement and one, in which its brain has evolved, for perception; its mouth is on its belly, and it has a circle of five eyes. On another world the dominant life form is a kind of organic ship, an immense sea animal with an intricate complex of bones and membranes which under muscular control function like sails. On some worlds there are highly evolved underwater civilizations, and on others myriads of insects or birds make up a single intelligent organism. On yet other worlds, where the pull of gravity is low and solar radiation high, life has evolved in the form of "plant men"—trees with

Below: R2D2 and C3PO, characters from the blockbuster science fiction movie *Star Wars*, directed by George Lucas. They are the latest in a long and illustrious line of movie robots ranging from the Tin Man of Frank Baum's *The Wizard of Oz* to the first female robot in Fritz Lang's *Metropolis*. The concept of androids (humanoid robots), though it stirs man's imagination, also expresses his fear that humans will be taken over by computers.

Some Bizarre Possibilities

Left: creatures of the Great Race which, according to American writer H. P. Lovecraft, inhabited earth for almost 150 million years. "All my stories," Lovecraft wrote, "unconnected as they may be, are based on the fundamental lore or legend that this world was inhabited at one time by another race who, in practicing black magic, lost their foothold and were expelled, yet live on outside, ever ready to take possession of this earth again."

eyes and limbs, leading a vegetable existence by day while drawing energy from the sun, by night moving around and engaging in a variety of social and cultural activities.

These images are reminiscent of the illustrations in children's books, where normally inanimate objects such as houses and trees possess limbs and sensory organs. Are such bizarre biologies possible? Scientists are divided on this question. "For all we know," writes Cornell University astronomer Carl Sagan, co-author of *Intelligent Life in the Universe* (1966), "biology is literally mundane and provincial, and we may be familiar with but one special case in a universe of diverse biologies." In *The Immense Journey*, published in 1973, Loren Eiseley writes: "Life . . . may exist out yonder in the dark. But high or low in nature, it will not wear the shape of man." On the other hand, some scientists have argued that man's shape is the most likely one for creatures of developed intelligence to have evolved, con-

"Humanoids on Other Planets"

sidering the physical conditions that must exist on any world for the existence of such intelligence. J. B. S. Haldane, the British biochemist, recalled in an essay a book he had had as a child which showed illustrations of giants ten times the size of a man. He calculated that, since they must have been ten times as wide and ten times as thick as well as ten times as tall, they must have weighed 80 to 90 tons, and their bone structure would have prevented such giants from taking a step without breaking their thighs. Haldane titled his essay "On Being the Right Size," and other writers have also argued that the right shape and form are essential preconditions for the evolution of an intelligent being.

This argument has been developed by Robert Bieri in his essay "Humanoids on Other Planets," first published in 1964 in the United States. Life, Bieri argues, has developed two distinct kinds of symmetry: *radial* and *bilateral*. Only stationary life forms have radial symmetry, which means they are balanced in relation to their centers of gravity like the spokes of a wheel. They also have a low degree of internal organization. Bilateral symmetry, on the other hand, is the balancing and streamlining of the two sides of the body, and it is essential to creatures that move, either in water or on land. Nervous systems of any complexity have only evolved in creatures with bilateral symmetry, and such creatures are diverse enough on our planet to show that bilateralism must be one of the first preconditions of the evolution of intelligence.

Bilateral creatures with complex nervous systems have other characteristics in common which are clearly not merely adaptive characteristics produced by chance, for they have evolved independently in many different species. They have at their front a mouth, sensing and grasping organs are located nearby, and the nerve center that controls these organs, the brain, is situated in the same area. This arrangement is obviously logical because, particularly in predators, sensing and grasping are essential to the gathering of food, and, in order to maximize speed and efficiency, the distance between the brain and the principal organs it controls should be as short as possible.

Birds, fish, and mammals all conform to this basic specification. But birds need to be light in order to fly and to maneuver in the air; they could not afford to carry the weight of a large brain along with the large heart and blood quantity required to supply it, so they would not be expected to evolve as intelligent creatures. The development of intelligence also seems to require three other factors: the use of tools, language, and social existence. Life in the water imposes severe limitations on all three of these factors, and so, Bieri argues, the evolutionary route to intelligence narrows down to a land animal of bilateral symmetry with its brain, mouth, and sense and grasping organs located at the front of its body. He next considers how the animal would move around. Since sliding and wriggling are rather slow, and biological tissue cannot bear up under the high pressures created by wheels, legs would seem the best means of locomotion. Jointed legs are adaptable to various types of terrain and provide maximum swiftness and efficiency. An odd number of legs would create a balance problem, and an excessive number would make for confusion and slowness, so Bieri concludes that not more than

Left: scene from the 1967 science fiction movie *Planet of the Apes*. The plot in this case involves speaking apes who rule over savage humans not yet able to think logically or speak. Darwin's evolutionary bombshell is given an ironical twist in the film when two ape scientists suggest during a trial that intelligent apes may have evolved from the contemptible humans. In the end it is revealed that the planet is actually earth many years in the future, after human society as we know it has destroyed itself through nuclear war.

six and not fewer than two pairs of legs would be best adapted to the needs of an intelligent land predator. Two pairs would be most consistent with the development of a large brain, because then one pair could be converted to arms for using tools.

Regarding sense organs, no doubt the creatures of another world would have evolved different degrees of capability for sensory response as well as other senses than those that human beings normally have, but on any world with conditions similar to earth they would surely have developed the five primary senses. Eyes of similar structure have evolved in a wide variety of unrelated creatures on our planet. Such eyes must evolve on any planet that receives light from a central sun. Natural selection favors two eyes because in a movable head they give the best all-around and depth vision—and information from more than two would confuse the brain. Two hearing organs would be expected for the same reasons, and they would evolve in any creature which lived in an environment where pressure changes might make an impression on its body. The same applies to the sense of touch. Sensors which could detect the presence of chemicals in the air or in water solutions would be naturally selected in an animal inhabiting a world where chemical reactions take place—that is, any world with life on it—and, to prevent poisoning, the animal's smell and taste sensors would have to be situated in or near the mouth.

This argument is based on what is known as the "theory of convergent evolution." It leads to the conclusion that intelligent beings of biological origin on other worlds would greatly resemble *Homo sapiens*, provided that physical conditions similar to those of earth prevailed. The argument further suggests that the best possible conditions for the development of high intelligence are those which prevail on earth, and that intelligence linked to biological life under any other conditions would necessarily mean a lower order than that of man.

This point can be illustrated by imagining what forms of life might have developed on other planets in our own solar system. Take first Mercury, the nearest planet to the sun. The surface

Other Planets!

Right: the planet Jupiter as photographed from the U.S. Pioneer 2 space probe in 1974. Largest planet in our solar system, Jupiter has a mass more than twice that of the other eight planets combined. Its surface is obscured by a turbulent, multicolored atmosphere thousands of miles thick.

Below: Jupiter's famous Red Spot, first observed in 1878. Although its nature and cause remain unknown, it has been observed to "float" relative to other identifiable points on the planet's surface.

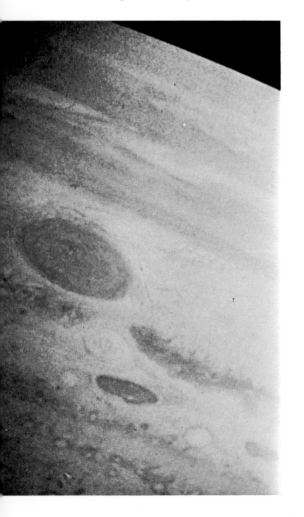

temperature on the sunlit side of Mercury is in the region of 700–750°F, which is unsuitable for the existence of any organism larger than bacteria. Lacking in water and oxygen, and having only a thin atmosphere of carbon dioxide, Mercury would be inhospitable to any form of life as we know it. Venus also has a surface temperature too hot to support life, as well as an atmospheric pressure 20 times that of earth. Venus is also surrounded by clouds 15 miles thick, which contain a 75-percent concentration of sulfuric acid. Mars might be more hospitable, but due to its low gravity, any evolved creatures would be tall and spindly. To survive in a climate with temperatures ranging from extreme heat to extreme cold, these creatures might have also had to evolve thick skins, which should be capable of changing color frequently in order either to reflect or absorb the sun's rays as required. On Jupiter the surface temperature and pressure would be too high to support life, but Carl Sagan has speculated that perhaps organisms similar to ballasted gas bags might float and feed at the different levels of the planet's atmosphere. He also points out that the gases of Jupiter's atmosphere "are just the components of the primitive atmosphere in which life arose on earth." The atmospheres of Saturn, Uranus, and Neptune are similar to Jupiter's, gaseous surfaces under high pressure, and any native evolved form of biological life would be inconceivable. The same applies to Pluto, the farthest planet from the sun, where temperatures must be too low to support life. If life exists at all in our solar system, it is most likely to be on one of the moons of Jupiter or Saturn, some of which may have a thin enough atmosphere, sufficiently low gravity, and tolerable though cold surface temperatures.

Advanced biological organisms are highly vulnerable to

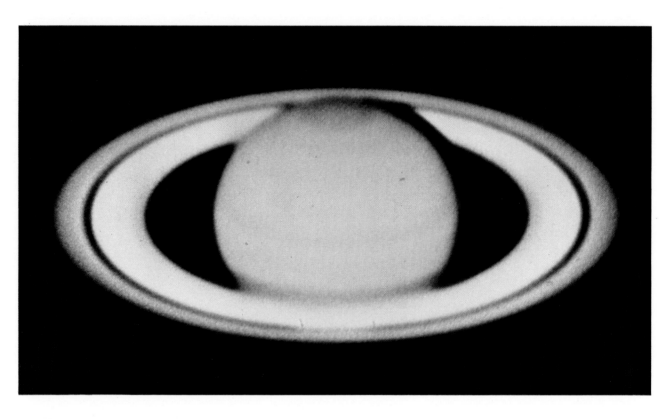

extremes of heat and cold, to excessive pressure, and to fatal contamination by radiation or poisonous chemicals. It is probably safe to say that advanced organisms could not have evolved on any other planet or planetary satellite in this solar system other than earth, except possibly on Mars millions of years ago. The possibility of creatures with a biology based on the element silicon, able to stand much greater heat than earth creatures with their carbon-based biology, is also rather small, despite the similarities between the chemical structures of carbon and silicon. Such creatures would have to breathe sand and excrete concrete, processes surely too uncomfortable and strenuous to permit much development of intelligence. Unless some form of disembodied biological life is conceivable, the physical and chemical laws would seem to limit the existence of any evolved form of life to earthlike planets in far distant solar systems. Convergent evolution theory suggests that any such evolved beings, if they existed, would not be fantastically different from man. They might be as different as, say, the dolphin, but the dolphin's streamlined shape is merely an adaptation to life in the water. Internally the dolphin is virtually identical to man—its organic structure and biology are the same.

One other possibility for the existence of biological life elsewhere is that put forward by an American mathematician, Freeman Dyson of the Institute of Advanced Studies at Princeton University in New Jersey. "Planets may be a good place for life to begin," Dyson has said, "but they are not a likely place for the home of a big technological society." Societies which are much more advanced technologically than ours (much of our technology, especially relating to space exploration, is but a few decades old) could have developed ways of doing what Dyson

Above: the planet Saturn. Circling the sun some 400 million miles beyond Jupiter, Saturn is an even colder and less hospitable world. Its unique ring system, made up of many billions of independently orbiting ice particles, measures some 169,000 miles from edge to edge, but it is probably no more than 10 miles thick.

Right: the planet Uranus. Third of the great planets, it is very similar to Saturn, though smaller and colder. Again it has a very low density, and its atmosphere contains methane and ammonia. It is circled by five satellites.

Below: the surface of the planet Mars, fourth planet from the sun. The terrain is as thickly cratered as that of the moon, a fact revealed by the U.S. Mariner 4 space probe of 1965. The larger craters—Nix Olympica, for example, is about 310 miles in diameter—tend to be flat inside, while the smaller ones have a less eroded appearance with sharper edges and curved floors.

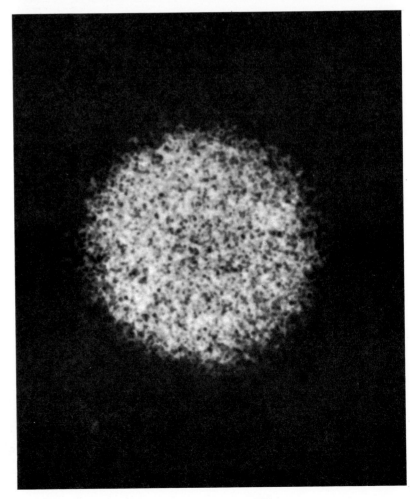

calls "astroengineering"—they could have colonized comets. "Countless millions of comets are out there, amply supplied with water, carbon, and nitrogen, the basic constituents of living cells," he claims, and it would certainly not be beyond the capability of an advanced technology to develop on comets the two essential requirements for human settlement: warmth and air. Dyson puts his theory forward as a vision of man's possible future, foreseeing the day when we learn to grow trees and other plants on comets, and introduce a variety of animals. We could create for ourselves "an environment as beautiful as ever existed on earth," perhaps even "starting a wave of life which will spread from comet to comet without end until we have achieved the greening of the Galaxy." Though this may be a bold and startling conception to us, that does not mean that it could not already have been achieved by beings who have been around longer and evolved further than man.

But is life necessarily and exclusively a biological phenomenon? In other words, is life as we know it the only form it can take? Perhaps it is appropriate to take further the question raised in the opening chapter: what is life? That chapter concluded with the research and ideas of Dr. Harold Saxton Burr, who maintained that the bias of modern biology towards a chemical interpretation of life had obscured the demonstrable fact that life is

Astroengineering

Left: the moon's crust. This photograph was taken by one of the U.S. Surveyor series of unmanned spacecraft sent to explore the lunar surface between 1966 and 1968. It is one of the first color pictures of the moon. The surface is inhospitable and crater-scarred, and it offers an environment where even the hardiest microorganisms would perish. Since its atmosphere is a near vacuum, its surface is exposed to constant bombardment by sterilizing solar radiation and meteorites.

Car as Creature!

fundamentally an electrical phenomenon. Biologists are increasingly using information gained by electrical measurements to probe the mysteries of life, but in general they assume that the electrical properties of living things are a result of biological activity. Burr's contention, on the other hand, is that the forces of electrical fields cause and control biological activity—what he calls the L-field comes before and is independent of its physical embodiment. His idea has been elaborated by a British author, Edward W. Russell, who in his book *Design for Destiny*, published in 1971, presents evidence that thought is a form of energy. He suggests that fields of thought, or T-fields, must exist which would control and change L-fields and through them physical and biological processes. The idea that "mind" exists independently of matter is not such an outrageous or improbable one to modern physicists as it was to the majority of scientists a century ago. Indeed, the British physicist and astronomer Sir Arthur Eddington, who died in 1944, stated that "the ultimate stuff of the universe is mind-stuff." If such men as Eddington, Burr, and Russell are right, then life may be not only a biological phenomenon, but something which, while expressing or manifesting itself through biological processes, is independent of them.

Planets, according to the intelligent Black Cloud of astronomer Sir Fred Hoyle's 1960 novel, are extreme outposts of life, and it is most unusual to find animals with technical skills inhabiting them. The Cloud explains that the gravitational forces on planets limit the size to which animals can grow, thereby limiting their brain size and scope. Furthermore, the scarcity of basic food chemicals on planets and the slowness of synthesizing them from the basic energy source of sunlight "leads to a tooth-and-claw existence in which it is difficult for the first glimmerings of intellect to gain a foothold in competition with bone and muscle." Man, says the Cloud, is a rarity among planetary life forms, and "by and large, one only expects intellectual life to exist in a diffuse gaseous medium, not on planets at all."

In *Solaris*, another work of science fiction, first published in

Right: interior of a space ship from the movie *Solaris*, made from the book by Stanislaw Lem. Pressurized spacesuits, air locks, and other aids would be necessary for man to exist in space where the force of gravity and the protective effects of the earth's atmosphere do not exist.

English by the brilliant Polish writer Stanislaw Lem in 1971, the author imagines the existence on another world of a "sentient ocean." Originally like the primordial earth, a "soup" of slow-reacting chemicals, he proposed that this planet belonged to a two-sun solar system. Therefore it followed an erratic orbit, and in order to survive and evolve, the "soup" or ocean had had to develop the ability to influence the planet's orbital path. "Thus it had reached in a single bound the stage of 'homeostatic ocean' without passing through all the stages of terrestrial evolution, bypassing the unicellular and multicellular phases, the vegetable and the animal, the development of a nervous and cerebral system." Carrying this idea further, he explains, "In other words, unlike terrestrial organisms, it had not taken hundreds of millions of years to adapt itself to its environment—culminating in the first representatives of a species endowed with reason—but dominated its environment immediately."

It would be understandable if, as science writer Paul Weiss suggests, alien beings got the impression that the automobile was the dominant life form on earth.

Above: the "birth" of cars on a factory assembly line.

Above left: cars on an airfield.

Left: a frequent occurrence—the violent "death" of automobiles in a crash on the road. Presumably the human "parasites" are also killed or dislodged with the destruction of their "host" being.

The Automobile as "Earthian"

Below: a sinister car, fitted with steel spikes, made up from parts of destroyed cars—a still from the film *The Cars that Ate Paris*. In the movie the townspeople of Paris, Australia set up a series of road accidents around their town in order to trap motorists; eventually the youth of the town use these now-lethal cars to destroy the town.

These science fiction ideas generated by the fertile minds of Fred Hoyle and Stanislaw Lem may seem far-fetched, but they are not idle fantasy. They raise and intelligently discuss the question whether life is a characteristic exclusive to biological systems, and they agree in their conclusion that it is not. Other evidence also points to the same conclusion. If we accept that nonbiological life exists, it might be appropriate to consider whether it could exist on our planet. An American science writer, Paul A. Weiss, in an essay published in 1964 entitled "Life on Earth (by a Martian)," has suggested in fun that if extraterrestrials visited earth they would assume the automobile was the dominant organism on the planet. They would probably regard human beings as its "obligatory parasites," which never stray too far away from and soon reenter their hosts, and which show "only extremely limited capacity for independent active motion." Weiss writes an imaginary mission report of the visitors, describing their observations of the brain (engine) of the "Earthians," their birth on assembly lines, and their occasional violent death in combat with each other. The report is a joke, of course, but it does make the point that biological organisms may not be the most obvious "life forms" on earth. It also suggests that human beings may be more dependent on their own manufactures than they realize. If this is true of the automobile, it is even more

Left: a lunar module transported to the moon by Apollo 11 being checked by U.S. astronaut Neil Armstrong. Space exploration equipment has been developed which is capable of drawing itself in (the characteristic of contractility)—considered by some scientists to be an essential attribute of living things.

applicable to the computer, the modern "thinking machine" which can perform many functions of which formerly only the human brain was capable.

A living organism has been defined by some scientists as having three essential characteristics: *irritability*—the ability to react to environmental influences; *conductivity*—the ability to pass information from one part of itself to another; and *contractility*—the ability to draw itself in. Modern computers certainly possess the first two of these properties, and space exploration technology has developed devices capable of the

Left: modern computers possess two other characteristics of living organisms— irritability and conductivity. The photograph shows the largest of a new family of computers.

"The Universe is Mind-Stuff"

Right: one of man's more lifelike creations. This traffic warning robot, christened Mr. Sam, waves a flag to warn oncoming motorists of street repairs ahead. In operation near the Arc de Triomphe in Paris, France, Mr. Sam is fitted with a polyester arm operated by a 12-volt battery.

Opposite: *Autumn Meeting* by Tim White. The confrontation between man and a superior nonbiological creature has been envisioned by many writers, artists, and filmmakers. The reactions on both sides usually vary from interested wonderment to hostile mistrust, but, as with many aspects of life elsewhere, to know what would really happen we must wait and see.

third. If we accept these three conditions as definitive, it could be said that man has created life. Whether the life he has created might ultimately take over is a question that has been raised by many writers, both in imaginative fiction and in rational non-fiction predictions of the future. There are functions of the human mind that machines could never simulate, but on the other hand it is equally true that there are aspects of machine thinking that human beings could not and probably would not wish to emulate. The assumption that a machine's programer necessarily understands and controls the development of his program is no more true than that a father understands and controls the genetic legacy he passes to his offspring. It is possible to conceive a scenario of the future in which intelligent machines, initially of man's making, later learn to reproduce themselves and become the dominant and most highly evolved life form on the planet earth. In such circumstances biological life, which has generally been considered the only possible form of life, might become obsolete and be regarded as merely an evolutionary stage, the means by which a Creator could bring forth his pride and joy— the best and most highly evolved living creature.

Chapter 13
The Ageing Process

Why must we get older every day, with our bodies eventually running down and sometimes becoming unable to perform the most basic functions? Can science offer a "cure," or at least slow down the speed of the decline? This depends, of course, on what causes ageing, and so far all that scientists know for certain are the symptoms. What relationship exists between diet and long life—is chronic starvation an alternative to early death? Why have so many great thinkers such as Voltaire and Bertrand Russell lived particularly long lives—is there a connection between intellectual activity and length of life? Will the fabled "fountain of youth" ever be found—or should we concentrate on living sensibly during whatever time is granted to us?

When King David was "old and stricken in years" and could not get warm although "they covered him with clothes," the Bible says he was advised to find a young virgin who would cherish him and lie with him. His servants found for him Abishag the Shunammite, which is why the belief that sleeping with young women can revitalize an ageing male is sometimes called shunamitism. It is, perhaps not surprisingly, an idea widely distributed throughout the world. There is an English folk saying that "while the miller grinds the old man's corn, the miller's daughter grinds the old man young again." Indian kings of the 11th century sought to prolong their lives by having intercourse with maidens specially trained for the purpose. Chinese Taoists taught that "he who is able to have coitus several tens of times in a single day and night without allowing his essence to escape will be cured of all maladies and will have his longevity extended; if in one night he changes his partner ten times, that is supremely excellent."

Marsilio Ficino, an Italian Renaissance philosopher and physician, prescribed that an old man should "find a young, healthy, gay, and beautiful girl, attach his mouth to her breast, and drink her milk while the moon is waxing." This recommendation obviously has its roots in crude magical thinking, and the very idea of shunamitism might be due to the lust and wishful thinking of the old. On the other hand, it has been observed that male rats in captivity live considerably longer if a young female

Opposite: every human body ages, a little bit every day, and so far as we know this is an unavoidable process. This old servant was photographed in the mid-1860s at the Victorian manor house where she probably lived and worked most of her long life. Mystics, religious believers, and faddists have tried various ways to prolong or regain youth, but their efforts have proved less than successful. The search for a "fountain of youth" continues today.

Old Husbands, Young Wives!

is put into the cage with them. Medically it could be suggested that the presence of a desirable female stimulates the production of the male hormone testosterone. The desirable female, however, need not necessarily be a beautiful virgin; she might be an old man's own old wife. But neither shunamitism nor continued marital sexuality can long or radically affect the processes of ageing to which every human body must eventually succumb.

Two centuries ago the American statesman and inventor Benjamin Franklin wrote that "all diseases may by sure means be prevented or cured, not excepting even that of old age, and our lives lengthened at pleasure even beyond the antediluvian standard." During the course of those two centuries, however, man's average life expectancy has not reached even the 84 years that Franklin himself managed to clock up. Medical science has made great progress in saving human life, but hardly any in extending it.

Observing that flies seemingly drowned in wine often revive when taken out, Franklin expressed the wish that when he was near death he should be immersed in madeira and then later revived. A similar but more scientifically sound idea is the modern practice of *cryobiology*: freezing a body just after death so it can later be revived and cured when medical science has become sufficiently advanced. There is no doubt that life processes can be suspended by subjecting them to extremely low

Right: *The Old Lover* by the 16th-century German painter and engraver Lucas Cranach. The concept of the rejuvenation of an old man through the caresses of a beautiful young woman—shunamitism— has enlivened the folk wisdom of many countries throughout the world. It may be true that the yearning to appear desirable to a younger woman can stimulate an old man's system and make it produce the hormones necessary for feeling and looking like a much younger man.

temperatures. Bacteria that have been frozen for thousands of years have been revived, and there are hundreds of children in the world today born of artificial insemination of a woman with sperm which have been stored by freezing. There is some doubt whether human beings can be frozen for long periods and expect to be revived with all their functions and faculties intact; despite this, by 1977 at least 13 people, some quite prominent members of society, had had themselves deep frozen and shut away, and the total must be considerably higher today.

Cryobiology, of course, is not a means of prolonging active human life. It can only offer a means of deferring death in the hope that some physician, probably as yet unborn, will have the skill and goodwill to reverse whatever biological processes have brought the poor frozen body to death's door. Precisely what those biological processes are is one of the mysteries of life, but they are gradually yielding to diligent and widespread research. At present there is still no scientific consensus as to which of several theories of *senescence*, or ageing, is closest to the truth.

The signs are there for all to see. As people grow older, their hair loses its pigment and goes gray, their skin loses water and becomes wrinkled, muscles lose their tone and become weaker resulting in flabbiness of the flesh, teeth rot, eyesight and hearing becomes less acute, connective tissues harden and contract so that the body tends to stoop, bones lose calcium and become

Left: *Married Couple* by the 20th-century German painter George Grosz. It sometimes seems that old men who still have the love and companionship of their wives don't show their age as much as men who are left alone. Perhaps the necessities of living with and relating to another person "keep the juices flowing" as much as would the advances of a younger woman.

Age and Death

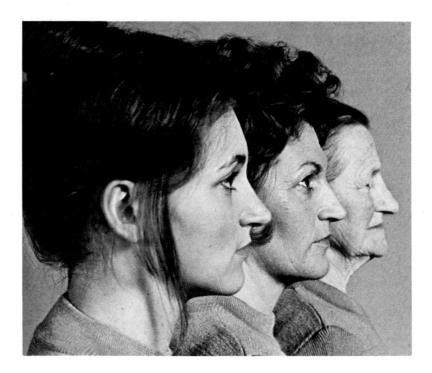

Right: three generations of women in one family. In an extended family where several generations form a closely knit unit, ageing may not seem so frightening since the resulting change in roles is a progression rather than an end. Older people have a part to play and experience to pass on, and where age is valued perhaps it is not dreaded.

more brittle, and joints are more likely to become crippled by arthritis. Internal processes also deteriorate: heart output drops, the efficiency of the liver and kidneys falls off, arteries become clogged, cells die off and weight loss is common, particularly in the brain, and the rate of cell replacement in vital organs is greatly reduced. The strength of a man's hand grip at 75 is 55 percent of what it was at the age of 30, his heart output is 70 percent, his maximum long-term work rate is 70 percent while for short bursts it is 40 percent, his taste buds have reduced in number by 36 percent, and his oxygen intake during exercise is 40 percent. These, as Shakespeare writes with bitterness, "are the rewards reserved for age, to set a crown upon our lifetime's efforts." Or, as the Duke in *Measure for Measure* puts it: "When thou art old and rich, thou hast neither heat, affection, limb, nor beauty, to make thy riches pleasant." Any of the world's literatures would yield a score of similar quotations. Human beings everywhere age and become decrepit; most don't like the fact and many look to science to do something about it.

The complexity of the problem of doing something about it is highlighted by the length of the list of processes and functions which are prone to degenerate. There is no single cause of ageing, and although people often say that someone "died of old age," this can never be true because old age is not a disease but a condition of increased likelihood of disease. Senescence, wrote Professor J. Z. Young in 1971, is "the gradual accumulation of defects." Dr. Alex Comfort, an American, writes in his book *Ageing: The Biology of Senescence* (1964): "Senescence shows itself as an increasing probability of death with increasing chronological age: the study of senescence is the study of the group of processes, different in different organisms, which lead to this increase in vulnerability."

The fact that there are differences in the ageing process be-

tween different organisms has suggested to some researchers that studying other organisms might provide a clue to human ageing. In the 1930s British scientist G. P. Bidder raised the question whether senescence was in fact a general law of life. He pointed out that there is no evidence that ageing takes place in certain reptiles or fish. Moreover, many forms of life such as trees, fish, and sponges are not limited to a specific size as mammals are, therefore there are no grounds to believe that indefinite growth is not natural. Clearly there must be something to regulate the rate of growth in man to stop it completely when he reaches his maximum size in his 20s. Bidder suggested that "senescence is the result of the continued action of the regulator after growth is stopped." His theory would appear to be supported by the fact that human beings do diminish in size with age, but it is scarcely adequate to explain all the other biological changes that take place.

Organisms that age at a much slower rate than mammals apparently do not undergo senescent changes at the cellular level. A cutting taken from an old tree will grow at the same rate as when it was young. Human body cells are much more highly differentiated than the cells of trees, plants, and lower vertebrates and may be more prone to error when they divide and multiply. It is therefore possible that organic systems age more rapidly the more highly evolved they are. Comfort has suggested that the halting of growth in vertebrates might be of two distinct kinds. When a cell population of a specific size and make-up is reached in the lower animals, it may remain static, governed by a simple process of replacement. In more evolved biological systems the process of differentiation may continue and the cells may be less stable as a result. The more highly differentiated cells, for instance those of the nervous and muscular systems, may be subject to "selective nonrenewal." We know from experience of mechanical

The ageing process in the oak tree. Above left: young oak seedling.

Above: a flourishing English oak tree.

Below: the Major Oak at Edwinstowe, Sherwood Forest. At an age of 410 to 420 years, the trunk has reached a maximum girth of over 32 feet.

"Faulty Copying"

Above: a one-week-old baby and (above right) an adult hedgehog. Man is the most long-lived of mammals, but he is also the only one reluctant to face the inescapable fact of ageing.

systems that greater complexity makes for greater vulnerability, and it seems logical that the same should apply to biological systems.

Comfort distinguishes three categories of biological material: cells that divide and multiply throughout life, cells which are incapable of division or renewal, and noncellular material which makes up connective tissue and forms, for instance, the arteries, lungs, tendons, ligaments, and intestinal walls. Each of these categories of material is vulnerable in a different way, requiring three basic hypotheses of the ageing mechanism: in the first category it is the result of changes or "faulty copying" in the multiplying cells, in the second it is attributable to injury or loss of nonmultiplying cells, and in the third it is caused by wear and tear and gradual changes in the connective tissues. The three hypotheses are not mutually exclusive. Most biologists would agree that senescence is generally a combination of the three types of degeneration, and that in particular individual cases one type dominates.

This three-category distinction makes it possible to specify some of the fundamental causes of the many senescent changes that occur. Each of the three general hypotheses, moreover, is an umbrella concept under the heading of which are more specific theories.

The theory that ageing and its ills and infirmities are basically

Queen Victoria as a young girl, a mature woman, and two years before her death at the age of 82.

Above: a water color of 1840 by A. Penley. Above left: Victoria presenting a Bible in the Audience Chamber at Windsor Castle around 1861.

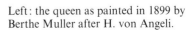

Left: the queen as painted in 1899 by Berthe Muller after H. von Angeli.

Signs of Ageing

due to changes at the cell level is known as the "somatic mutation" theory. It is supported by observable facts. For instance, the number of cells which have more than the normal 46 chromosomes increases with age, and the children of older parents tend to be shorter-lived and are more likely to be abnormal than those of younger parents. That multiplying cells are prone to "memory loss" after a number of generations is suggested by the Hayflick limit. It is also known that faulty copying of the DNA molecule in the process of cell division, as well as the resulting genetic mutation, are increasingly likely to occur the older an organism gets. The fact that radiation, both from natural and man-made sources, affects mutation rate suggests that it may be a factor in the ageing process. A minority of biologists even believe it is a factor of major importance.

Sir Peter Medawar, the British zoologist, has put forward a plausible theory to explain why there is a sharp rise in fatal diseases such as cancer in the later part of life. Although an individual's genes do not normally change, different genes become effective at different stages of life. Over numerous generations the struggle for survival would have tended to eliminate the harmful genes that express themselves early in life, leaving those that express themselves later unaffected because carriers

The 17th-century Dutch painter Rembrandt van Rijn was a keen observer of himself as well as of his fellow human beings, as evidenced by several careful self-portraits. (National Gallery, London.) Below: the artist at the age of 34. Below right: Rembrandt in the last year of his life.

of them would have already reproduced before their effects became obvious. Medawar suggests that as men learned to avoid the dangers of death by accident that make the life expectancy of animals relatively short, they would increasingly suffer from the effects of those later-operating fatal genes which had not been eliminated. This theory certainly matches the observable facts— that the incidence of cancer rises sharply in that portion of the population over 50 years old and is rarely traceable to a specific cause. But it offers little to those who want a secret of longevity, except the rather impractical suggestion that they should be born of long-lived parents.

The second general hypothesis, that ageing is caused by loss or injury of nonmultiplying cells, is also supported by observable and measurable facts. Some of the body's irreplaceable cells are lost daily as a human being grows older, and the eventual effect is loss of both body and brain weight. Whether loss of weight necessarily implies loss of function is debatable, however, particularly with regard to the brain. On the one hand, there can be full recovery of functions after brain damage, while on the other hand brain and nerve cells die when they have no function. Also, as Young points out, cell death occurs at all ages and is not simply a symptom of degeneration. "There are systems in the developing organism that ensure the death of a proportion of the cells . . . to lead to the continued life of the whole." The rate of cell loss cannot be tied simply to the rate of ageing.

Injury and damage to nonmultiplying cells are probably more specific causes of senescence than simple cell loss. Ageing cells are increasingly vulnerable to errors in the protein synthesis mechanism and in their enzyme systems, and, of course, once such errors have occurred they will not be eliminated by cell division. Cells also appear to lose with age their ability to identify and keep out toxic elements, and there is also the ever-present hazard of injury induced by radiation.

The third category of senescent changes is those which affect the noncellular material of the body. Some 30 percent of the body is made up of connective tissue, which is composed of two basic proteins, *collagen* and *elastin*. With age collagen, a fibrous substance, becomes rigid, and elastin loses its elasticity. These changes become obvious in the wrinkling of the skin and are also responsible for hardening of the arteries and the general loss of flexibility and suppleness. Many biologists think that what is called "cross-linking" of collagen is the prime cause of ageing. The long spiral atom chains that make up the protein macromolecules have sensitive spots on them that can easily link up with other molecules or atoms, and, as increasing numbers of such cross-links are formed, the collagen becomes more rigid. A number of chemical compounds in the body function as linking agents between proteins, and calcium is the most abundant of these. Overeating is believed to precipitate ageing because it increases the rate of cross-linking, and it would appear that life can be prolonged to a certain extent by avoidance of overeating. Biochemists seeking a formula for an "anti-ageing pill" have concentrated their research on finding something that will inhibit cross-linking and even unlink cross-linked molecules. This is currently one of the most promising lines of research.

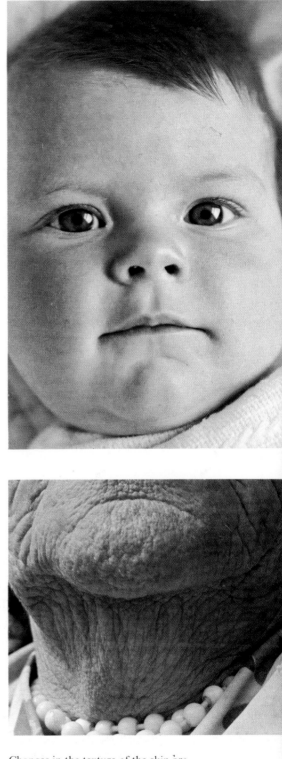

Changes in the texture of the skin are frequently due to age.

Above: the skin of a baby is soft and elastic, while that of an elderly person is rough, wrinkled, and sagging. This is because the protein collagen becomes rigid with age.

Above: the bleak and solitary Finnish
landscape of North Karelia near the
Russian border. Here in 1971 a special
health research project was set up to study
why local men aged between 25 and 60 are
1.5 times as likely to die of strokes and
heart attacks and almost twice as likely to
die of throat and lung cancer as men living
in the big city of Helsinki.

To these three general principles of the biology of ageing
should be added a fourth which takes more account of external
influences. Two basic factors have a considerable influence:
temperature and nutrition. Young observes that in Florida
lizards only live for one year, whereas in the colder weather of
Maryland, where they are less active, they live for four years. He
proposes that the rate of ageing is related to the rate of energy
turnover. Comfort makes the same point when he writes that
"something, which, for want of a better term, we have called
'program,' must run out and be succeeded by senescence."

This is the principle in operation when someone "works himself to death" or dies young because he has been "living it up."
Energy turnover tends to be more rapid in a hot environment,
but for mammals it is also rapid in a very cold environment,
because they must increase their metabolism to keep up their
body heat. The British physicist Anthony Lawton has calculated
that if the human body could function at a temperature of 84°
instead of 98°F, its lifespan could extend to 140 years, but it is
doubtful whether such a change would be desirable since all our
actions, including our rate of thinking, would be proportionately
slowed down.

Ageing and Life Style

Left: Orthodox crosses in a local cemetery bear witness to the extraordinarily high death rate among young men who, considering the physical exercise they get in clear, open air, should presumably have been among the fittest on earth. Possible causes of this phenomenon are the high rate of cigarette smoking, the intense cold, and a diet of foods high in animal fats. Severe drinking binges and extensive use of refined white sugar have also been suggested as reasons.

The most obvious external factors influencing biological processes are the things we eat and drink and the air we breathe. Biologists generally agree that one contributory cause of ageing and the illnesses of the old is the cumulative effect of absorbed and inhaled poisons. There is a much higher mortality rate from lung cancer among people who live in an urban metropolis than among those in rural areas, and the heavy cigarette smoker (considered as smoking more than 20 per day) has an average life expectancy of eight years less than the nonsmoker. Regarding diet, numerous theories as well as proven facts show that certain foods tend to precipitate ageing and fatal diseases and others tend to extend life expectancy. It is clear that the most readily available means of prolonging life is through diet control.

Many people eat yogurt because they believe that Bulgarian peasants have enjoyed uncommonly long and healthy lives because of it. The idea was put forward in the 19th century by the Russian bacteriologist Elie Metchnikoff, who believed that ageing was the result of cumulative self-poisoning through toxins produced by bacteria which live in human intestines. Supposedly the process could be counteracted by replacing the resident bacteria with lactic acid bacilli present in yogurt. The theory has few supporters today, but yogurt remains a popular "health food," and justifiably so, though for other reasons than Metchnikoff proposed.

The soundest dietary theories relating to ageing are those which tell us what not to eat. Animal fats and oils and refined sugar and its products certainly have harmful long-term effects, and foods that tend to increase calcium concentration in the arteries, such as eggs, should be eaten sparingly in middle and later life. But if experiments with animals are anything to go by, it is not so much what we eat as how much and when that affects our longevity, and the real secret of long life may be systematic starvation.

Experiments with rats have shown that their lifespan can be doubled by slowing down growth in early life through restricted

The elderly folk of Upper Sheringham, Norfolk, in eastern England, hold the British record for longevity. A London pathologist, Dr. David Davies, believes the reason is the rich mixture of trace elements such as iron, calcium, chromium, and selenium in the local soil from which much of the fresh produce comes. The statistics are that 15 percent of the villagers live beyond the age of 75, while the average in the rest of north Norfolk is 11 percent and only 5 percent in Britain as a whole.

Right: Fred Dennis, 72, still rides his bicycle to his job as a gardener.

Below: Elsie Livesey (left), 80, the oldest living woman born in Upper Sheringham, and Eva Loades, 70, who came to the village 45 years ago.

calorie intake. Some have been kept in a state of retardation for up to three years before being put on a normal diet, whereupon they rapidly matured and lived a life not only long but also free of the infirmities and chronic diseases that often afflict older rats. In other experiments, increases in life expectancy of 20 percent have been achieved by starving the animals for one day in three, and similar results have been noted with a moderate reduction of general food intake. In these cases, too, the incidence of disease was significantly reduced. Experiments with mice have produced the same results, and there can be little doubt that the same must hold true for humans.

While some researchers have been trying to retard senescence in rodents, others have been seeking to accelerate it. They have found that the most effective way to do this, apart from exposure to radiation, is by means of induced neurosis. Here is another lesson for man. It was taught by the 18th-century British social theorist William Godwin, who suggested that clear thinking, cheerfulness, and emotional stability made for longevity, while confusion, worry, and negative emotions were likely to cause premature death. Godwin himself lived into his 80s; many great men famous for their reason, emotional balance, and intellectual power have equaled or excelled him—the British philosopher Thomas Hobbes, who lived to be 91, Voltaire, the French philosopher and man of letters (84), Irish dramatist George Bernard Shaw (96), and the British philosopher Bertrand Russell (94). Psychologists have studied the relation between intellectual activity and survival in the old and shown clearly that the degree of mental vigor is related to the length of life. This is not really surprising, as it is clear from the principles of psychosomatic medicine that mental states profoundly influence body states. The will to live and a sense of purpose can often be the crucial

Psychology and Your Longevity!

Left: the 17th-century British philosopher Thomas Hobbes, painted by John Michael Wright about 10 years before his death at the age of 91. It has been suggested that a vigorous mind is a factor in human longevity.

Left: a drawing in pastels by Nicholas Cochin of the 18th-century French philosopher and writer Voltaire and his niece. Voltaire, who lived to be 84 years old, was one of the leading figures of the Age of Rationalism.

factors that decide whether a critically ill person lives or dies, and they can sometimes even act to reverse cancers which doctors have declared incurable.

Professor Young has suggested that a relatively recent development in human evolution might have been the natural selection of longer life "because of the value of the information store to a larger group." In other words, the capacity for wisdom, advice, and know-how that the old have and could contribute to their society or family may be useful characteristics in terms of survival. It is true that, for all his worry over his mortality, man is the longest-lived of land mammals, and Young's theory may well account for this fact. The theory would further imply that a man's chance of longevity will be related to the degree of social activity and importance, the amount of attention and respect from others, and the number of interests of his own that he continues to enjoy late in life. It has often been observed that many men accustomed to work, responsibility, respect, and power die soon after retirement if their subsequent lives do not give them these satisfactions or others to compensate for them.

In the science fiction novel *Trouble with Lichen*, which British author John Wyndham published in 1960, a scientist drops a speck of lichen into a saucer of milk and then notices that the lichen seems to prevent the milk from turning sour. From this observation an "antigerone," or antidote to ageing, is developed. The research team find themselves holders of a drug which is not only inestimably valuable in a commercial sense, but which also confronts them with the ethical problems surrounding the power

Above: the British philosopher and mathematician Bertrand Russell at a literary lunch. Russell lived to be 94, and up to the very end of his life in 1970 he was actively involved in campaigns for peace and compassion in a war-torn world.

Right: the Japanese Emperor Hirohito at an annual rice-planting ceremony in 1976. Born in 1901 and a survivor of his country's defeat by the Allies in World War II, the emperor attributes his fitness to taking daily cold baths.

Mind against Age

Left: Morarji Desai, born in 1896 and elected Prime Minister of India in March 1977. His diet consists of fruit, nuts, and milk, and he has remained celibate. He is one of the many millions of Indians who by family choice, religion, or tradition are strict vegetarians. Some not only reject all forms of flesh but also some vegetables. Abstention and moderation (he does not drink alcohol) are part of his strict discipline.

to grant or withhold the prize of prolonged life. The story is a fascinating investigation of the possible psychological and social repercussions of the discovery, although it is doubtful that the finding of such a miraculous cure or remedy for ageing is likely in the foreseeable future.

The dream of finding a "fountain of youth" or an antidote to ageing has always attracted people, and very often some have deluded themselves or others into believing it has been found. The medieval alchemists believed that gold prolonged life. The reason that Chinese Taoists believed in the retention of sperm may have been because they had observed that some creatures, salmon for example, often die soon after spawning. Several years ago a number of prominent elderly people were given injections of "monkey gland," which did make them look and feel younger for a while since it was basically a hormone treatment, but which did not noticeably prolong their lives. Today the fashion is for pills, and a variety of "youth" or "anti-ageing" pills have been developed. Dr. Richard Passwater of the American Gerontological Research Laboratory has for years been testing a number of these, and he has some hopes of success with a pill containing vitamin E, an antioxidant, a protein resynthesizer, and an ingredient which protects against radiation.

Dr. Passwater's research, unlike most previous attempts, is at least based on the realization that there are multiple causes for ageing. However, it is still doubtful whether even a composite medication of this kind can be developed in the foreseeable future which could arrest or delay the complex process of senescence. What may conceivably happen as a result of both current research and the application of the knowledge we already possess is a reduction in the frequency of diseases of age, a longer period of vigorous and active life, and possibly an extension of average life expectancy by some five or ten years. But those who look to science or to magic for some cure-all, while neglecting to take sensible precautions to insure a longer, healthier, happier, and more significant life for themselves, will probably always be disappointed.

The Fountain of Youth

The Fountain of Youth by the 16th-century German artist Lucas Cranach. The dream of finding an antidote to age (and possibly a means to eternal life) has attracted people for centuries. Doctors, chemists, and alchemists as well as medicine men and priests have searched, obsessively in some cases, for a place, a combination of ingredients, or a method of stopping the ageing process. But perhaps the most successful individuals are those who, accepting that they will not remain young, manage to contribute something to their society and enjoy a happier life than those others who spend their allotted years clawing at a bygone youth.

Chapter 14
What is Death?

Just how common is it for people to be buried alive? Bizarre cases have been recorded of corpses coming back to life, either before burial or, less fortuitously, afterwards. Can these people be described as "recovering from a severe attack of death"? The accepted signs of death, or symptoms of the disease as some would claim, are not particularly reliable, and the needs of modern transplant surgery put great pressure on doctors to determine death quickly, before the entire body dies. Can our technology cope with these problems? What suggestions do current psychical researchers offer for the age-old mystery of what happens after death? Their conclusions may prove unexpected.

In 1964 a New York surgeon died of shock when a "dead" man on whom he had just begun an autopsy suddenly rose and seized him by the throat. He paid a high price for an error that has been made by many medical men before him. Misdiagnosis of death has always been one of the physician's occupational hazards, and until modern instrumental techniques were introduced to confirm death many people must have been buried alive. Indeed, many horror tales are recorded which tell of coffins opened because sounds were heard coming from them—the unfortunate occupant usually found dead but in a position which showed there had been a struggle, sometimes with bruises and lacerations to indicate the desperate panic to get out. In the 19th century, fear of premature burial was such that one ingenious inventor marketed a device which would enable someone who regained consciousness in his grave to activate a multiple alarm system including a bell, a flag, and a flashing light, and to breathe air through a tube until rescuers arrived.

When a doctor signs a death certificate he is protected by the wording that the diagnosis is made "to the best of his knowledge and belief." If he has attended the patient during the fatal illness he does not actually need to examine the body before issuing the certificate. A case was recently reported of a doctor who was informed by a child that his grandfather had died. He went to the grandfather's house, saw that he was lying in bed apparently dead, wrote out the certificate, and left it on the bedside table.

Opposite: *Sir Charles Aston at the Death Bed of his Wife* by the 17th-century painter John Souch. Death has always been frightening and mysterious as well as full of grief for those left behind. Bewilderment, on the other hand, is perhaps the uppermost emotion in the victim's mind.

Right: frontispiece of *The Danger of Premature Interment* by Joseph Taylor, published in 1816. It shows "the body of Tulliola, the daughter of Cicero, discovered entire and uncorrupted in a sepulcher, 1500 years after burial." It is extremely unlikely that a body would remain "entire and uncorrupted" for such a long period, but the fear of being buried alive was a very real one for many people. A quote from the Taylor book reads, "To revive nailed up in a coffin! A return of Life in Darkness, Distraction, and Despair! The Brain can scarce sustain the reflection, in our coolest moments."

Some time later he phoned the coroner's office, who sent someone to the house. The official was in the bedroom reading the death certificate when Grandpa sat up and asked him who he was and what he was doing there. The man thought quickly enough to say he was a police officer and had entered the house to investigate because the front door was open. Grandpa soon recovered from his "death" and six months later was active and well.

One of the most bizarre cases of recovery was recorded by a French professor of medical law. He told of a young monk who stopped one night at a house where a young girl had recently "died" and was laid out for burial. He offered to keep watch over the corpse, and during the night he stripped the girl's body and had intercourse with it. He went on his way the next morning and preparations were continued for the girl's burial, but as she was about to be interred she recovered, and nine months later she bore a child.

To speak of someone recovering from death sounds odd, but as the South African biologist Lyall Watson suggests in his book *The Romeo Error*, published in 1974, it makes sense to regard

Life and Death

death as a disease and even to speak of being "only slightly dead" or "very seriously dead." It is a disease with quite specific symptoms: cessation of breathing and heartbeat, drop in body temperature, changes in the eye, skin pallor, rigidity of the muscles, and lack of reflexes. These are the signs of death as listed in standard medical textbooks. The combination of all of them does indicate fairly reliably that all vital functions have permanently ceased, but no one symptom in itself is an infallible sign of death. Cases have been known of people showing all of these symptoms for a time and yet "coming back to life." It is reasonable to say of such people that they suffered from a severe attack of death but fortunately recovered.

Each of the symptoms deserves more close consideration. Cessation of breathing as a sign of death has long been known to be fallible. Putting a feather over someone's lips to see if it moves or holding a mirror to his mouth to see if it mists up are tricks used in Shakespeare's day and are still used today as rough-and-ready ways of testing for life, but they are notoriously unreliable. A man can train himself to reduce his oxygen intake drastically, and Indian fakirs have proved that they can survive for hours with very little by having themselves buried in airtight boxes. English medical texts of the past cite the case of a Colonel Townsend, who demonstrated in the presence of three doctors that he could deliberately stop breathing for as long as half an hour. The doctors found when they tested him that his pulse had apparently stopped as well, and they were about to leave him for dead when he reversed the process and slowly returned to normal.

Apparent cessation of pulse and heartbeat is another highly unreliable symptom. Human beings can survive with very slow heartbeats—modern heart surgery is conducted at such low temperatures that the organ is effectively stopped for the duration of the operation. Again, Indian yogis have demonstrated while wired to an electrocardiograph that they can stop their hearts at will.

When police pathologists are asked to estimate how long a person has been dead, the main indicator they use is body temperature, which falls by a factor of 2°F for each hour for the first 12 hours after death. Although temperature drop is useful in this particular context, it is not always reliable. Death by some diseases, such as cholera, actually results in a post-mortem rise in temperature, until with the process of decomposition the body temperature returns to normal for a time. People trapped by avalanches or blizzards have been known to survive with body temperatures as low as 63°F (normal is 98.6°F), and in winter conditions old people have often been rescued from unheated rooms and revived when their body temperatures have fallen by more than 50°F.

Many medicolegal writers of the past considered that changes in the eye were the most reliable sign of death. Dilation of the pupil, drying of the cornea with the resultant haziness, and eventually the sinking of the eyeball all occur in stages through the hours immediately after death. However, they are not absolutely reliable signs, particularly in the early stages, for brain injury and some forms of poisoning can also produce, confusingly, the same symptoms as with the onset of death.

Above: the deaths of Romeo and Juliet in a drawing by Peter von Cornelius. The young lovers were tricked by the effects of a potion—it induced symptoms similar to those of death—and, like many before and after them, Romeo and Juliet were fooled.

Changes in the blood occur within hours of death which cause the well-known skin pallor of the dead. These include the clotting and settling of red blood cells under the pull of gravity, leaving a clear liquid near the surface of the skin in the blood vessels. But chemicals that prevent blood from clotting during life, which are produced in the cells lining the blood vessels, can remain active for days after clinical death has occurred, and their activity can eventually make the blood fluid again.

Rigor mortis, the stiffening of the body, normally begins six hours after death in the muscles of the eyelids and the jaw, gradually spreading to all the muscles of the body. It continues for about 12 hours and then gradually disappears, so that about 36 hours after death it is gone. The stiffening is caused by changes in the protein molecules of the muscles, so that if a person has violently exerted himself immediately before death it will set in very quickly, whereas if death occurs when a person is relaxed,

Deep Freeze Life

Cryonics is the freezing of human beings soon after death so that they can be reanimated and cured when science has progressed far enough. The first man to be cryogenized was an American professor of psychology, in 1967.

Left: the transport capsule constructed by Anatole Dolinoff, president of the Société Cryonics de France.

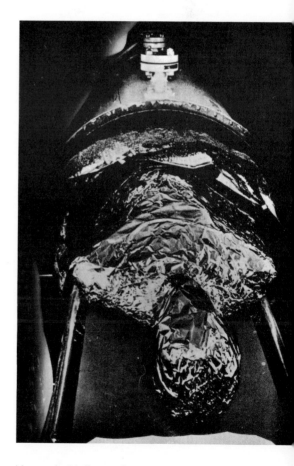

Above: the bodies await resuscitation in a stainless steel cryotorium in a vertical position, constantly supplied with liquid azote and kept at a temperature of around -320°F. Scientists meeting in 1963 predicted that man would attain immortality by the year 2100.

such as after a long illness, it may be slight. A cold environment will delay it while a warm one will hasten it. The short duration and variation of its occurrence in general make it unreliable as a means of diagnosing death. Cessation of reflexes, the last of the classical criteria of death, can also occur in catatonic and trance states, so that, like the other criteria, on its own it is totally unreliable as a means of determining whether a person is alive or dead.

Since each individual sign of death is unreliable, no combination of them can be said to be 100 percent certain. Modern techniques for measuring brain wave activity using an electroencephalograph are less fallible, but people in a state of chronic narcosis due to barbiturate poisoning have been observed to register an EEG similar to that of death for several hours and then revive. Overdoses of drugs can, in fact, result in a convincing combination of death symptoms, as can drowning, and because of the great increase in modern times of attempted suicide by these means the problems involved in certifying death from any cause have begun to concern doctors and medicolegal authorities more and more.

Both professional and public concern over the exact definition of death have increased with the advent of transplant surgery. The problems that physicians once had with certifying death could always be circumvented by the commonsense method of "wait-and-see," for when decomposition set in after about three days there could be no doubt that the death process was complete. But today's doctor, who must certify death because a vital organ is to be donated in order to save or extend another person's life, cannot wait and see. In fact, the organ must be alive and functioning when it is transplanted, so that death must be certified on the basis of some other principle than the permanent cessation of all vital functions. Thus a distinction between "heart death" and "brain death" has come about in recent years.

Normally heart death precedes brain death, for it is the cessation of the heartbeat that stops the blood from carrying to the

The Transplant Phenomenon

A heart transplant operation performed by the American surgeon Dr. Arthur Cooley. Right: a scene from the operation by Dr. Cooley at a hospital in Houston, Texas.

Below: Dr. Cooley examines the donor's heart before transplanting it. The final stages of the operation must be completed swiftly, because the sooner the new heart can take over from the heart-and-lung machine the better.

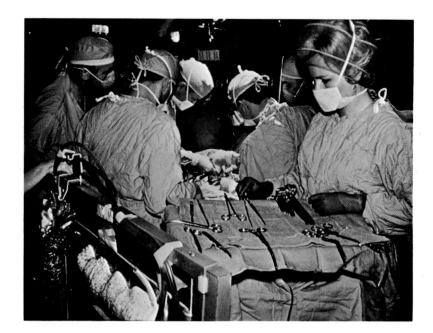

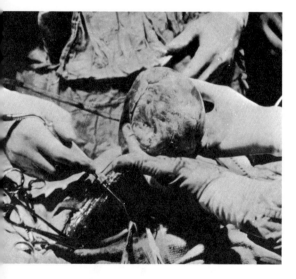

Right: survivor of a separate heart transplant operation. Henri Debord, 46 years old, recuperates in a sterilized room some days after the operation.

brain the oxygen and sugar it needs to survive. The brain can extend its survival for some time after the heart has stopped by drawing on its reserves. But with the development of modern techniques for mechanically maintaining heartbeat, brain death can precede heart death in the hospital situation. Brain death is irreversible, and when it occurs the essential attributes of personality disappear. It is therefore possible to claim that a person whose brain has ceased to function is no longer alive, and that it is justifiable to keep the heart or kidneys functioning artificially if they can eventually be grafted onto another patient and thereby save a life. Most organ transplant donors are accident victims, and often they have suffered brain injury. Some countries try by law to safeguard against premature certification of death by requiring that two doctors, one of senior status and neither connected with the transplant team, must sign the death certificate. Sometimes life support systems must be turned off before a

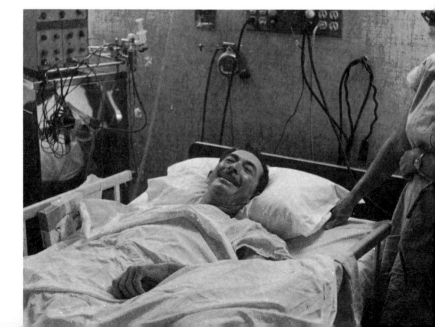

person can be pronounced dead, and in these cases the doctors must have established not only that the EEG reading is consistently flat but also that oxygen consumption in the brain has ceased.

It has only become possible to distinguish between heart death and brain death since electrical monitoring devices were developed, but other distinctions have long been known. A manual of medical jurisprudence published in 1836 states, "Individuals who are apparently destroyed in a sudden manner, by certain wounds, diseases or even decapitation, are not really dead, but are only in conditions incompatible with the persistence of life." The distinction may seem comical and hair-splitting, but it does draw attention to the important point that after "sudden death" many life processes continue in the body for several days. As the British biophysicist Joseph Hoffman wrote in his book *The Life and Death of Cells* (1958), "From the standpoint of the atomic arrangement of living matter, death brings no abrupt change of affairs." Gradually the living system as a whole runs down, its cellular parts cannot maintain their organization, energy is used up and not replaced, and the body is reduced to a collection of low-energy chemical substances.

The phenomenon of cell death, because it is a continuous process essential to life, makes it more difficult to define death in general. Physicians can, within the limitations described, define clinical death and specify the point at which a human being has ceased to be viable. But if cellular as opposed to human life is considered, the process is characterized by continuous cycles of destruction and creation in which death is an inseparable part of life. Could it be that human existence is similar, and that the bodily life of any individual is a relatively insignificant part of an ongoing process of which the individual has no conception?

"Each cell," writes Hoffman, "seems to have in its make-up the blueprint for its special job and, what is perhaps more important, directions for its death. In this respect cells are like clocks wound up to live through a specific span of time in a limited region of space. In other words, cell death is programed. Recent studies by Dr. Jerome Wadinski at Brandeis University in

Above: two-year-old Leslie Sturgis, believed to be the youngest child ever to receive a transplanted adult kidney. She is shown with her mother and Dr. Oscar Salvatierra, who performed the operation at the University of California's Moffitt Hospital. The transplant kidney was donated by a 17-year-old male victim of an automobile accident.

Left: a dead octopus. Studies at Brandeis University in Massachusetts suggest that the death of the female octopus may be programed in her cells from birth. It is not yet clear exactly how this "death hormone" works or whether a similar substance exists in other animals.

Massachusetts suggests that the death of the octopus is programed, and this research has led to a fascinating discovery of how the program may be altered. Normally the female octopus dies very soon after her eggs have hatched, and, according to Dr. Wadinski's work, this may be caused by a secretion from a small, round, yellow gland, known as the optic gland, located near the brain. When the nerve to this gland was experimentally cut in some females, it was found that they laid eggs prematurely, when still very young, and often did not die after the eggs had hatched. This observation led to further experiments in which the gland was completely removed after the female had laid her eggs, and the result was that its removal gave the female octopus a new lease on life.

This research is very recent, and it is still not clear whether the "death hormone" works by a simple process of stopping the octopus from feeding or whether it causes self-poisoning; studies of its chemistry are currently underway at Cambridge University in England and elsewhere. Nor do we know whether similar programing takes place in human beings, although, of course, if it does it would not be related directly to the reproductive function as it is in the octopus. However, it is certainly true in general that if cells fail to die at their appointed time to make way for

Right: *The Apotheosis of Nelson* by the late-18th- and early-19th-century American painter Benjamin West. The 18th-century British naval hero's final battle was Trafalgar in 1805. His ship the *Victory* broke through French lines, but he himself was killed by a sharpshooter in the last stages of the victorious action. The death of a hero has inspired visions of heavenly glory since the time of the Greeks.

"Death Hormone"

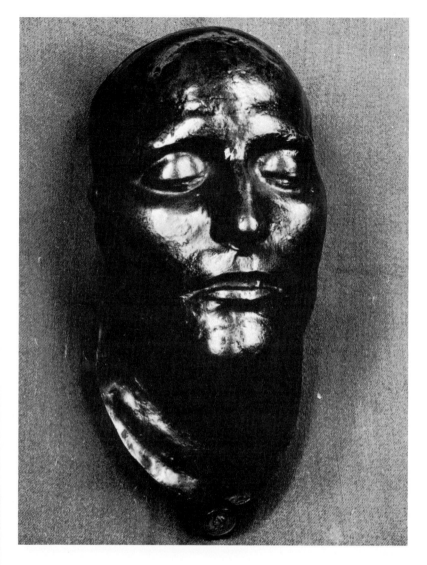

Left: the death mask of Napoleon Bonaparte. Casts have been taken from the faces of the dead for centuries in many different cultures. It is perhaps a manifestation of the reverence in which the dead are frequently held as well as a gesture of defiance of the physical obliteration of death.

new ones, they may form dangerous growths such as cysts, tumors, or cancers, which can cause the death of the entire organism. Life is dependent on programed cell death.

Scavenger cells also exist in the bloodstream which prey upon other cells—ingesting bacteria and microscopic particles—as predatory animals prey upon others. Again, the life and health of the organism as a whole depends on their activity. There are also many body cells which, in the interests of the body's survival, are required to give up their individuality and their ability to multiply, such as the shinglelike cells of the outer surface of the skin, the glasslike cells of the lens of the eye, and the cells in the root of the hair shaft which dismantle their nucleus and become part of the hair. The body sheds millions of these cells every day—"enough of them to fill a soup plate," Lyall Watson estimates—and if the process were halted the weight of the body would theoretically double in about 40 days. In actuality, however, before there could be much weight increase the death of the body would be caused by the blood vessels becoming clogged with excess cells.

So, as Hoffman writes, "every moment of life for the animal

body depends upon the certain death of cells in its tissues. How this most important death for tissue cells is brought about is one of the basic mysteries of the living process." Cells that die return to the pool of matter from which all life is continually created, and the following lines by an ancient Sufi poet represent a fundamental truth of the biochemistry of life:

> "I died as a mineral and became a plant.
> I died as a plant and rose to animal.
> I died as animal and I was man.
> Why should I fear? When was I less by dying?"

But though it is true biochemically speaking that there is no "final" death, most human beings do not feel the contentment of the poet. Many people see themselves as something more than their biochemistry, and they fear and resent the fact that their individual personality and consciousness will after a limited time apparently be snuffed out.

But is this true? The ultimate mystery of life and death is whether biological death necessarily implies utter and final extinction of personality. There is evidence that life may not be an exclusively biological phenomenon, and if this is true then there is a distinct possibility that a nonphysical aspect of human personality exists which survives the death of the body.

Many people who have been very close to death or have actually "died" in the sense that their vital functions have ceased for a period of time, and who have then recovered, subsequently proclaim their unshakeable conviction that neither personal identity nor important mental abilities would be extinguished by

Below: *The Bridge Across*, a psychical drawing by a medium. People have always wondered what follows the moment of death, and many suggestions have been put forward. Spiritualists are considered by many to be reliable guides and reporters of what "life after death" might be like.

physical death. An American medical doctor, A. S. Wiltse, described his "death" as follows:

"Feeling a sense of drowsiness come over me, I straightened my stiffened legs, got my arms over my breast, and soon sank into utter unconsciousness. I passed about four hours in all without pulse or perceptible heartbeat, as I am informed by Dr. S. H. Raynes, who was the only physician present. During that time I came again into a state of conscious existence and discovered that I was still in the body but the body and I had no longer any interests in common."

Dr. Wiltse's account continues, "With all the interest of a physician, I beheld the wonders of my bodily anatomy, intimately interwoven with which, even tissue for tissue, was I, the living soul of that dead body. I watched the interesting process of the separation of soul and body."

When the process of separation was complete Dr. Wiltse found that he was standing in the room among other people, gazing at the body on the bed. It was lying, he said, "just as I had taken so much pains to place it." He tried to get the attention of the others. He bowed playfully, saluted, and laughed, as he thought, very loudly, but nobody seemed to notice. He reflected, "They see only with the eyes of the body. They cannot see spirits. They are watching what they think is I, but they are mistaken. That is not I. This is I and I am as much alive as ever." When eventually Dr. Wiltse returned, or, as he felt, was sent back to his physical body, he felt "astonishment and disappointment" to find himself alive again in the conventional sense.

"I am as Much Alive as Ever"

Below: *The Plains of Heaven* by the early-19th-century British painter John Martin, known for his wild imaginative power. Artists have visualized heaven in many ways, from the biblical Christian concept of the Renaissance to the ambiguous feelings of modern times. One of the continuing mysteries of life is the fate awaiting humans after death.

The Dying and the Laying Out

Below: *The Last Dream* from a Victorian monument by J. Edwards to the late Miss Hutton of Sowber Hill near Northallerton in Yorkshire, England. The concept of the soul as a separate entity which leaves the body after death was an important one to the Victorians. The verse below this monument reads: "Her sun went down while it was yet day / But unto the upright there ariseth light in the darkness."

Right: a headhunting ritual traditionally accompanied an Otsjanep funeral in West Irian, Indonesia. Headhunting was associated with increasing the earth's fertility and with the victim's obligations as a servant in the next world. The head was closely associated with the soul, hence the relationship between decapitation and the preparation of the body after death.

Disappointment at having to return to life is expressed by many other people who have had pseudo-death experiences. Their accounts have many features in common with the Wiltse case, which is by no means unique. These accounts also share features with the various traditions and teachings on the death experience which have been passed down through scriptures and oral traditions in cultures throughout the world. These have been set down in, for instance, the Egyptian *Book of the Dead*, the Tibetan *Bardo Thodol*, and the Chinese *Secret of the Golden Flower*, ancient texts which include detailed instructions for guiding the soul through transitions to higher planes of existence.

When the British novelist D. H. Lawrence was dying in 1930 he described to his friend Aldous Huxley, who was with him through his last hours, how he experienced the separation of his "double" from his physical body and saw it standing in a corner

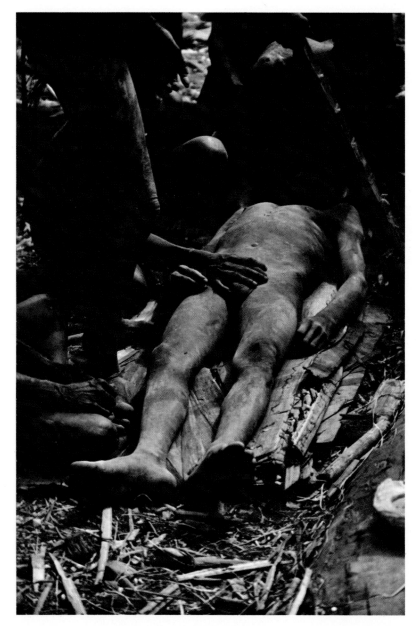

of the room looking back at him. Other people who have attended the dying have themselves seen a vaporous emanation from the body at the moment of death, and in England in 1907 Dr. Duncan McDougall, a physician, made the curious observation that the human body loses about 2.5 ounces in weight at the moment of death. Unable to explain this weight loss by any hypothesis of physical change, he asked, "Is it the soul substance?"

Sir William Barrett, the British physicist and psychical researcher who died in 1925, related the case of a "Mrs. B" in his book *Deathbed Visions*. While she herself lay near death in an

Above: a painting on papyrus from the ancient Egyptian Book of the Dead of Userhetmos. It dates from around 1320–1200 B.C. The Egyptians were extremely aware of death; some of the pharoahs seem to have been more interested in their tombs than in the events of their lifetimes.

"I Cannot See"

English hospital, the death of her sister Vida had been kept from Mrs. B. for some weeks in order to avoid distressing her. In her last moments Mrs. B. murmured: "It is all so dark, I cannot see." Her last words before she died, spoken with an expression of puzzlement, were, "He has Vida with him . . . Vida is with him."

A number of cases are recorded in which a dying person has spoken of meeting someone of whose death he or she had no normal means of knowing. For the dying to speak of seeing friends or relatives whom they knew had died before them is even more common. The research director of the American Society for Psychical Research, Dr. Karlis Osis, sent out questionnaires

Right: D. H. Lawrence, 20th-century British writer, in a painting of 1920 by Jan Juta. Lawrence spoke on his deathbed of experiencing the separation of his double from his physical body.

Left: the 20th-century British writer Aldous Huxley, who sat with his dying friend D. H. Lawrence trying to understand what was happening.

Above: Sir William Barrett, British physicist and psychical researcher, author of *Deathbed Visions*. His wife later edited *Personality Survives Death: Messages from Sir William Barrett*. Barrett was responsible for documenting "reports from the other side" of death.

to 10,000 nurses and doctors concerning the behavior they had observed in dying patients. Out of the 35,000 deaths of which he received accounts, just 10 percent of the patients had been conscious and rational at the end. More than 700 of these died in an "extremely elevated" mood, many speaking in their last moments of familiar helpers who had come to welcome them. These could, of course, have been wishful hallucinations. An explanation to cover the less usual case, when the patient speaks of seeing someone whose death should be unknown to him, may be the working of extrasensory perception. Although we cannot know the truth for certain, the facts are suggestive.

So are some of the facts of mediumship. Even allowing for the possibility that many so-called mediums may be charlatans, and others may be unconscious telepaths capable of drawing information from the minds of other people, there are still many cases in modern history of mediums whose feats are not apparently explainable by either of these hypotheses. Consider the following example.

The late 19th-century Australian psychical researcher Richard Hodgson was highly skeptical of claims of survival after death, but after investigating the Boston medium Mrs. Piper for 15 years he became convinced of it. One morning while on his way to a sitting with Mrs. Piper he read in the newspaper of the death of a person whom he later referred to as "F." He knew that this person had been a close relative of a Madame Elise, a friend of his who had died some time before and who had several times

Below: *Christ Glorified in the Court of Heaven* by the 15th-century Italian painter Fra Angelico. After his death and resurrection Christ rose to heaven, where he was acclaimed the Son of God by a host of saints and angels. In traditional religions death marks the beginning of a new and better life, the life of the soul, as a reward for true believers. Will we ever know for certain that there is life after death?

communicated with him through Mrs. Piper. Madame Elise spoke to him at that day's séance and told Hodgson that F. was with her but not yet able to communicate. She also said that she had been with F. when he died, and she told Hodgson what she had said to him at that time. Some days after this sitting with Mrs. Piper, Hodgson heard an account of F.'s death from a completely different source. He was told that before F. died he had claimed that Madame Elise was speaking to him, and he had repeated her words. These words, which Hodgson said were "an unusual form of expression," were precisely those that Madame Elise had repeated to him through Mrs. Piper.

Such cases of apparent communication with the deceased— and there are many others, equally if not more puzzling as well as being irrefutably authenticated—suggest that to regard a human

being as merely the sum of complex biochemical processes is too limited a view. It reflects the restrictions of old-fashioned scientific rational materialism, which still prevails in Western thought although it could be considered incompatible with the findings of modern physics. Another view of human personality, claiming that it comprises a nonphysical component which separates from the physical body at death and survives it, has been held by most of the world's religions from time immemorial. It would appear to be supported by a wealth of evidence from modern psychical research. It may be impossible to make an unprejudiced choice between the two views, and for a final answer to the question "What is death?" we shall just, as the 20th-century British philosopher C. D. Broad says, have to wait and see, or perhaps wait and not see.

The Christian Consolation

Index

References to illustrations are shown in italics.

Picture Credits

Key to picture positions: (T) top (C) center (B) bottom; and in combinations, e.g. (TR) top right (BL) bottom left

Aerofilms 211(TL); Photos Paul Klinger, courtesy Professor Bernard Agranoff 190; J. A. Dyal, *Transfer of Behavioural Bias and Learning Enhancement: A Critique of Specificity Experiments*, Akademiai Kiado, Budapest, 1971 189; Aldus Archives 12, 29, 39(T), 40(T), 68, 70(B), 80, 91, 166(R), 168(T), 236; © Aldus Books 51, (Rudolph Britto) 39(B), (David Cox) 19, 23, 31(B), (Gordon Cramp) 54(L), 130(B), (Design Practitioners) 22(L), (Jon Kenny) 65(L), (Eileen Tweedy) 203, (Sidney W. Woods) 38, 42, 64(T), 65(R), 130(T), 134(B); Hugo Iltis, *Life of Mendel*, George Allen & Unwin Ltd., London, 1966 30; John Napier, *Roots of Mankind*, George Allen & Unwin Ltd., London, 1971 97; Heather Angel 18, 21, 113, 126(T), 131(R), 135, 137(L), 142, 152(B), 153–154, 221–222, 241(B); Hubertus Strugheld, *Your Body Clock*, Angus & Robertson (U.K.) Ltd., London, 1972 149; Associated Press Ltd. 209; Barnaby's Picture Library 49, 50(B), 115, 213(T); S. Summerhays/Biofotos 246(R); Birmingham City Art Gallery 122; Black Star 134(T); Reproduced by permission of the Trustees of the British Museum 191; British Museum (Natural History) 47(L), 88(R), 90, 171; California Institute of Technology 31(T), 35; Karl Lashley, "In Search of the Engram" from *Symposia of the Society for Experimental Biology, IV*, Cambridge University Press, 1950 187(L); Camera Press 33, 50(T), 57, 72–73, 99, 106(T), 226–227, (IMS/Hallberg) 211(B), (Marion Kaplan) 98; *Canadian Journal of Neurological Sciences*, Winnipeg, May, 1976 186(L); Carnegie Institution of Washington 69(R), 71; Photos FCP © Aldus Books, courtesy of the manufacturers, Catalin Limited 16, 17(T); The Chemical Society 13, 56(L); Bruce Coleman Ltd. 22(R), 106(B), (Des Bartlett) 140, (S. C. Bisserot) 94, (R. & M. Borland) 100(R), (Jane Burton) 133, 157(T), (Stephen Dalton) 136, (Oxford Scientific Films) 132, (Masod Qureshi) 114, (Hans Reinhard) 157(B), (H. Rivarola) 131(T), (Norman Tomalin) 102(R); Robert Ardrey, *The Social Contract*, William Collins Sons & Co. Ltd., London, 1970 112; Musée des Beaux-Arts, Paris/Photo Conzett & Huber 8; Cooper Bridgeman (British Library) 162, (National Gallery, Prague) 159; Gene Cox (Micro Colour International) 26, 48; Released in the United Kingdom by Crawford Films Ltd. 212; Dr. Tim Crow 172; *Daily Telegraph* Colour Library 60(T), 70(T); Professor José Delgado 164–165, 173–174; Colette Portal, *The First Cry*, J. M. Dent & Sons Ltd., London, 1972 and Alfred A. Knopf, Inc., New York 69(L); From Dragon's World Ltd. (© Bruce Pennington) 196, (© Tim White) 201, 215; E.M.I. Medical Limited 184; *Proceedings of the IRE*, November, 1940, © by The Institute of Electrical and Electronic Engineers, Inc. 126(B); Reproduced by Gracious Permission of Her Majesty Queen Elizabeth II 67; *Eltern Magazine*, Hamburg 220; Fitzwilliam Museum, Cambridge 161; Werner Forman Archive, 46(L), 55, 247; William C. Dement, *Some Must Watch While Some Must Sleep*, W. H. Freeman and Company Publishers, San Francisco, 1974 146; Giraudon 166(L); Harvard University Archives 187(R); Courtesy of the Historical Society of Pennsylvania 111(T); Michael Holford Library photos 110, 242; Honeywell Information Systems Ltd. 213(B); IBM United Kingdom Ltd. 193; Imperial War Museum, London 60(B); India Office Library and Records 20, 150; Institute of Geological Sciences 14–15(B), 198; Institute of Neurology

40(B), 186(R); Keystone 32, 37, 144, 214; Courtesy Professor Dr. Ivo Kohler 128–129; Kunsthistorisches Museum, Vienna 44, 158; E. D. Adrian, "Sensory Areas in the Cerebrum of the Pig," *The Lancet*, July 10, 1943 137(R); Heinz Hoflinger/Frank W. Lane 101(T); Electron micrograph by Gordon F. Leedale 17(B); J. F. Lehmanns Verlag, München 169(T); Lady Lever Art Gallery, Port Sunlight 107; Linnaean Society 10–11(B), 76, 81–82, 83(T), 88(L); Sir William Barrett, *Personality Survives Death*, Longmans Green & Co. Ltd., London, 1937 249(R); Josiah Macy Jr. Foundation, New York 169(B); Manchester City Art Gallery 234; The Mansell Collection, London 28, 46(R), 52–53, 85(R), 108, 151, 170(L), 246(L), 249(L); Marshall Cavendish (Alan Duns) 176–177, (Patrick Thurston) 225; Barcelona Museum/Photo Mas 218; The Menninger Foundation, Topeka/Photos Don Richards 180–184; NASA 147, 199, 206–208; National Film Archive 200; National Gallery, London 66, 121, 224, 250–251; National Library of Medicine, Bethesda, Maryland 89; National Museums of Kenya 92–93; National Portrait Gallery, London 167, 223, 229(T), 248, (Courtesy of *Punch*) 61(L); New York Historical Society 229(B); *The Osteopathic Physician*, October, 1972 138; Courtesy Professor Ian Oswald, Edinburgh University 160; Bo Jarner © Pressehuset/PAF International 101(B); David Paramor Collection, Newmarket 61(R); Parker Gallery, London/Photo © Aldus Books 104; The Peale Museum, Baltimore. Gift of Mrs. Harry White, in memory of her husband 83(B); Taken from *Mechanics of the Mind*, Cambridge University Press, based on results of Penfield and Bolding 168(B); Dr. A. C. Allison, *New Biology*, Penguin Books Ltd. 54(C)(R); Pfizer Inc., New York 178; Courtesy Professor D. C. Phillips, F.R.S., Oxford University 15(T); Popperfoto 103, 105, 111(B), 119–120, 230(B), 240(T)(C), 241(T); *Radio Times* Hulton Picture Library 56(R), 77; Photo Réunion des Musées Nationaux 148(B); Rex Features Ltd. 230(T), 231, 239, 240(B), (Gilles Carron) 102(L); Darwin Museum, Down House, courtesy of Royal College of Surgeons/Photos Eileen Tweedy © Aldus Books 84, 85(L), 86–87; Royal Society 43; Reproduced by courtesy of The Marquess of Salisbury 58–59; Scala 78, 156; Science Museum, London 10(T), 34, 36, 170(R); after John B. Calhoun, *Population Density and Social Pathology*, February, 1962. © by Scientific American, Inc. All rights reserved 116–117; Michael S. Gazzaniga, *The Split Brain in Man*, August, 1967. © by Scientific American, Inc. All rights reserved 194–195; Photomicrograph Landrum B. Shettles 64(B); Society for Psychical Research/Mary Evans Picture Library 244; Dr. Arnold H. Sparrow/Photo Robert E. Smith, Brookhaven National Laboratory 47(R); H. S. Burr, *Blueprint for Immortality*, Neville Spearman Ltd., Sudbury, Suffolk, 1972 25(B); Photo Staatliche Museen, Berlin-Dahlem 233; Städelisches Kunstinstitut 238; *Sunday Times* Colour Library 124–125; Syndication International 228; The Tate Gallery, London 192, 219, 245; Musée de l'Homme, Paris/Photo courtesy Thames & Hudson Ltd. 74; Curt Paul Richter, *Biological Clocks in Medicine and Psychiatry*, Charles C. Thomas, Publisher, Springfield, Ill. 145, 155; Transworld Feature Syndicate 62, (Mike Peters) 100(L); Courtesy Twentieth Century Fox 202, 205; Nathaniel Kleitman, *Sleep and Wakefulness*, University of Chicago Press, 1963 148(T), 152(T); University of Oslo/Photo O. Vaering 2–3; VW Motors Ltd. 211(TR); Victoria & Albert Museum, London (Photo Cooper-Bridgeman) 243, (Photo Eileen Tweedy © Aldus Books) 216; Visual Programme Systems Ltd. 210; Yale University Art Gallery 25(T).